W9-CEO-739

PLANTS & GARDENS

BROOKLYN BOTANIC GARDEN RECORD

INDOOR BONSAI

FIRST PRINTING 1990
SECOND PRINTING 1991
THIRD PRINTING 1992

Brooklyn Botanic Garden

STAFF FOR THIS ISSUE:

SIGMUND DREILINGER, GUEST EDITOR

BARBARA B. PESCH, DIRECTOR OF PUBLICATIONS

JANET MARINELLI, ASSOCIATE EDITOR

JO KEIM, EDITORIAL CONSULTANT

AND THE EDITORIAL COMMITTEE OF THE BROOKLYN BOTANIC GARDEN

BEKKA LINDSTROM, ART DIRECTOR

JUDITH D. ZUK, PRESIDENT, BROOKLYN BOTANIC GARDEN

ELIZABETH SCHOLTZ, DIRECTOR EMERITUS, BROOKLYN BOTANIC GARDEN

STEPHEN K-M. TIM, VICE PRESIDENT, SCIENCE & PUBLICATIONS

COVER PHOTOGRAPH BY CHRISTINE M. DOUGLAS
PRINTED AT SCIENCE PRESS, EPHRATA, PENNSYLVANIA

Plants & Gardens, Brooklyn Botanic Garden Record (ISSN 0362-5850) is published quarterly at 1000 Washington Ave., Brooklyn, N.Y. 11225, by the **Brooklyn Botanic Garden, Inc.** Second-class-postage paid at Brooklyn, N.Y., and at additional mailing offices. Subscription included in Botanic Garden membership dues ($25.00 per year).
Copyright © 1990 by the Brooklyn Botanic Garden, Inc.

ISBN: 0-945352-58-1

PLANTS & GARDENS

BROOKLYN BOTANIC GARDEN RECORD

INDOOR BONSAI

VOL. 46, NO.3, AUTUMN 1990

HANDBOOK # 125

LETTER FROM THE BROOKLYN BOTANIC GARDEN

ndoor bonsai began as a purely American idea, and BBG's former bonsaiman, Frank Okamura, was largely responsible for its general acceptance in the last two decades.

Bonsai courses have been conducted at the Botanic Garden since 1952, and in those early days students came from great distances to attend them. The plants provided were the classic Japanese species of pine, *Chamaecyparis*, maple, azalea and Zelkova. Gradually others were introduced. It soon became clear to the instructors that many local students were apartment dwellers who did not have outdoor cold-frames for overwintering hardy trees and that it was necessary to give them tropical or subtropical woody plants — which would survive central heating — to work on.

In the meanwhile, Frank Okamura was experimenting with many different plants: *Malpighia coccigera*, pomegranate, *Serissa foetida*, Citrus species, *Pyracantha*, boxwood and a hundred others. Eventually his collection of indoor bonsai varieties was large enough for a major exhibit. The two-week show attracted several thousand people during a snowy February in 1976.

The original handbook, guest edited later in 1976 by the late Constance Derderian (a "regular" in BBG courses), was a direct result of the evident interest in indoor bonsai. It has gone through 21 printings since then, and reached countless thousands of bonsai enthusiasts in many parts of the world.

Ficus benjamina 'Exotica', Exotic Fig Tree.

This year, BBG's Editorial Committee decided to publish an entirely new Handbook with articles by creators of tropical houseplant bonsai from many parts of the United States and other countries. Sigmund Dreilinger, another of BBG's early students, kindly consented to act as Guest Editor. Through his connections as President of Bonsai Clubs International he has been able to solicit articles from many knowledgeable bonsai growers. We thank Mr. Dreilinger and the contributors and hope you will be challenged to create your own bonsai after reading the words of those who have been doing so for many years, thanks to Mr. Okamura's leadership.

ELIZABETH SCHOLTZ
DIRECTOR EMERITUS

BASIC CONSIDERATIONS FOR GROWING INDOOR BONSAI

PHIL TACKTILL

An indoor bonsai is any woody plant trained as bonsai and capable of surviving the rigors of the indoor environment. In order to grow bonsai indoors, we must first understand the different plants used for bonsai and what they require to survive in their natural habitat. It then remains to alter the indoor environment to approximate the natural growing environment as closely as possible. The following is a list of factors one should consider:

Types of Materials

- **TROPICAL BONSAI:** plants requiring a warm environment to survive; unable to withstand a freeze.
- **BORDERLINE HARDY:** plants unable to withstand long periods of cold.
- **BROAD RANGE:** plants able to withstand extremes of hot and cold.
- **WINTER HARDY:** plants requiring a cold

PHIL TACKTILL, *owner of Jiu San Bonsai Co. in Farmingdale, N.Y., has been teaching bonsai classes for 20 years and has lectured and written extensively on bonsai.*

environment for a prolonged period of dormancy.

There are many charts available showing growing zones for different plants. Be aware that they refer to plants growing in the ground, not in containers. The roots on many trees are not able to withstand cold, nor the trunk and branches above ground. In bonsai containers, the exposure is multiplied and must be taken into account when planning which materials will be grown or maintained in which environment.

Almost all the aforementioned materials can be grown indoors if we provide the proper conditions. Here are some suggestions on how to provide a microenvironment suited to individual needs of the plants and their owners:

Increasing Humidity

- Spray foliage with an antitranspirant such as Wiltpruf, Envy or Evergreen. This provides a direct greenhouse effect on the foliage and is particularly useful after repotting to reduce moisture loss through the leaf surface.

- Set the bonsai container on a tray of gravel and water. The evaporation of water provides local humidity to the plant. The drainage holes in the bonsai container must be above the water level in the tray so that normal drainage of the soil is unimpeded.
- Use a humidifier. A cool mist vaporizer is also a good source of humidity in a smaller room, as is the standard warm mist vaporizer if the indoor temperature is kept quite cool (below 65°) in winter.
- Enclose the area in plastic. Any method of stretching plastic over a frame of wood, plastic tubing, or metal strips is workable indoors. Small tents for single trees or a collection of small bonsai, to larger indoor areas can be very successful as housing for tropicals.

Providing Air Movement

You can provide air movement with one or more fans, depending on the size of the indoor growing area. Or open a

INDOOR	**OUTDOOR**
HUMIDITY:	
Extremely low due to air-conditioning or heating (10 to 20 %). The Sahara Desert is about 30%.	In high coastal area, humidity is 40 to 80 %.
AIR MOVEMENT:	
Poor movement. Insects and fungi thrive.	Usually good air movement—often variable.
LIGHT:	
Poor at best.	Good to full sun.
TEMPERATURE:	
Stable, controllable.	Variable with extremes.

window, providing the outdoor air is not too cold. This is a good method in the in-between times of the year when the daytime temperatures are warming up but the nights are too cold for plants. Bonsai should never be exposed to drafts directly in front of a window in winter. Plastic can be used as a barrier if storm windows are not used or are ineffectual in keeping out even slight frigid drafts.

Regulating Temperature

Temperature can be altered to meet bonsai needs. When higher temperatures are needed, a thermostatically controlled space heater can be used. Sources of dry heat such as this should almost always be used in conjunction with a humidifier or other source of moisture in the air.

Those who grow indoors in warmer climates can use the refrigerator as a source of cold for trees needing a dormant period. Keep in mind that dormant trees do not need light and can be enclosed in the darkness of a refrigerator. Refrigerators remove moisture, however, and it is a good idea to spray the tree(s) with a fungicide and enclose it in plastic before refrigerating it.

Providing Light

Place bonsai near a window or in a greenhouse.

Provide artificial light with fluorescent cool white or daylight bulbs for sixteen to eighteen hours a day.

A 4' x 8 3/4' sheet of marine plywood (#6), reinforced with a 2" x 3" or 2" x 4" board rimmed with a 1/2" square strip

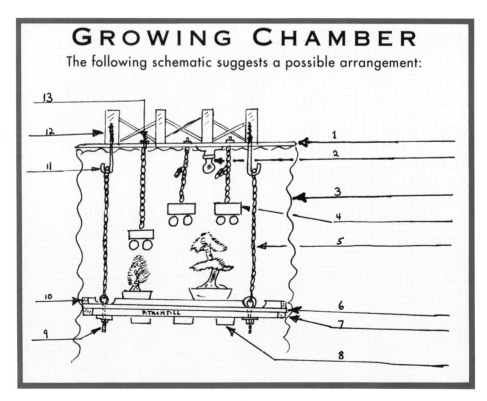

GROWING CHAMBER
The following schematic suggests a possible arrangement:

13
12
11
1
2
3
4
5
ATTACH FULL
10
6
9
7
8

Malpighia glabra bearing both flowers and fruit, from the Guest Editor's collection. PHOTO BY SIGMUND DREILINGER

(#7 and 10) forms the floor of the chamber. Paint the surface with epoxy, marine or urethane paint, preferably white to reflect light. This will preserve the wood.

A lighting fixture chain (#5) can be hooked to an eye bolt with nut and washer (#8 and 39), and the top hook can be screwed into the rafters (#ll, 12 and 13). Lamps can be suspended by chains so that they can be raised or lowered (#4). Incandescent bulbs can be mounted to the ceiling (#2). One incandescent bulb should be used for each sixteen feet of fluorescent tube. Three fluorescent tubes can be accommodated in the ceiling of the chamber, or suspended by chains from a 2x3 or 2x4 nailed to the rafters (#1). The lamp chains are mounted to the 2x4 with eye screws or bolts. The ceiling of the chamber should also be painted white to maximize reflection of light.

A plastic drop cloth should be used to enclose the assembly (#3). The base or ballast of the chamber, being remote from the fixtures and suspended, allows for additional circulation underneath the chamber.

These suggestions are basic considerations for supplying indoor bonsai with essential environmental needs. Common sense applied to individual circumstances enable the grower to make necessary modifications, and to invent specific methods to suit his or her lifestyle. 🎋

REPRINTED WITH PERMISSION FROM *WORLD TROPICAL BONSAI FORUM*.

ILLUSTRATED GUIDE TO STYLES

FRANK OKAMURA

Informal upright (Moyogi) style
with Shari branch
(dry part within foliage).

Slanting style with gnarled Shari
trunk (aged dead wood).

Slanting style with Jin.

FRANK OKAMURA *is former Bonsaiman at BBG. He was responsible for BBG's bonsai collection and is well known in bonsai circles.*

Forest style. Illusion of distance by
sloping. Suggestion of mountains.

Two-group style. Suggestion of two
human families.

Clinging upright style with
driftwood or stone.
Imagine a tree holding onto
a rock in a gorge.

Group slanting style.

One Grower's Tips for Success with Indoor Bonsai

JACK WIKLE

PHOTOS BY AUTHOR

Buxus microphylla 'Compacta' grown under fluorescent light for nine years.

Cotoneaster microphylla, ten years old, under lights for seven years.

My personal experiments in growing bonsai under lights began tentatively. Following the books' recommendations to keep the plants as close to the fluorescent light source as possible, I lowered the shop light over my basement workbench to six or eight inches above its surface. Then I brought in two small firethorns (*Pyracantha*) from the garden, pruned them hard, potted both in tiny bonsai pots and installed an economical timer set to turn the fluorescent light on 16 hours a day. This was the beginning of what has become a rewarding experience—growing and enjoying fluorescent light bonsai.

JACK WIKLE, *M.S. in Ornamental Horticulture, author of articles on bonsai and other subjects, lives and works in Michigan.*

Actually, it has seemed embarrassingly easy, far easier than outdoor bonsai culture so subject to temperature extremes, light intensity fluctuations and vagaries of wind and rainfall. I want to insist that if I can do it, you can too. But I must say that some people have rewarding experiences following my suggestions while others give up in frustration.

It's important to emphasize that even indoor bonsai require daily attention. This is not just being alert to insect infestations but also shifts in vigor or other evidence of change in the plant's well-being.

Many new indoor bonsai growers also fail to give the newly potted or repotted plant special attention during its recovery phase. This means watering well immediately after potting, then setting

13

the pot on a folded towel or sand bed to absorb excess water, thus letting more air into the soil mix. Roots require oxygen; every living cell must receive oxygen to survive. Even roots that are immersed in water must have oxygen to take water up efficiently. And the soil physicists tell us that oxygen diffuses some 10,000 times faster through air than it does through water. This is why a plant wilts in overly wet soil.

Special attention also means enclosing the newly potted tree in a transparent polyethylene bag—such as a food storage Baggie—to create an enclosure with 100% humidity in which the tree cannot dry out as it is adjusting. This works very well under fluorescent light but don't put the bagged plant in direct sunlight as heat buildup will quickly fry everything.

Another word of caution: The plant accustomed to the soft life in a high humidity chamber has many adjustments to make when the bag is removed and life in the outside world begins. To avoid severe damage or death, a "programmed re-entry" works well. The first day out, remove the bag for 15 minutes or a half-hour at most. Then each succeeding day double the time. (If you lack a dependable "sitter" to care for your fluorescent light bonsai while you are on vacation, the same bagging technique can be temporarily used. Water each plant well, enclose it completely in a polyethylene bag and set it back under the fluorescent light. I've done this for two weeks. Follow the programmed re-entry procedure again when you return—30 min., 1 hr., 2 hrs., etc. Also see E.O. Moulin's article, page 36)

After the first watering of the newly potted bonsai (by setting it in water up to the lip of the pot), it will not need any more water if kept from drying in a polyethylene bag for five to ten days as it

begins its recovery. It is when it is finally out of the bag that the most difficult water decisions occur. Since the plant is weak and root distribution is probably erratic, some of the water in the soil may be unavailable. So it's best not to wait for the soil to become visibly dry in this case. It is a difficult balancing act. You want the soil to dry—to increase oxygen availability—but not too much. As one student reacted, "I see what you're saying. You let it go until it's dry then water the day before." Yes.

After the tree has clearly recovered from the potting or repotting operation, close observation of wetting and drying patterns continues to be important. The tree that won't wait another day must be watered. But equally important, it's best to withhold water to prevent overmoist soil if it seems a tree could go another day without damage from drying. Although some plants are quite content in constantly moist soil, many others are actually healthier and grow better if allowed to dry between waterings. Once more, roots require oxygen as well as water and since plants have no circulatory system like we do for transporting oxygen, roots must take it up from air in the soil around them.

Watering the newly potted bonsai without washing away most of the fresh, and thereby unstablized, soil can be a problem. One approach is to use a misting bottle that produces a very fine spray. Another, and the one I favor, is to use a basting syringe for quick watering but precise control, thus avoiding erosion of the soil mix.

After the plant is well established and the soil mix has firmed, my watering technique is "quick dip and flush." I wet the soil mass by immersing it in plain water, then follow up immediately by pitcher-watering with a weak fertilizer solution. You can't find a pitcher with a fine enough spout to use with small bonsai? I couldn't either. Then I realized I could adapt the common, plastic household watering pitcher to my needs by epoxying a small button to the tip of the spout.

One reason for running fertilizer solution through the soil is that there can be problems without regular leaching (flushing out) of minerals the plant can't use. Salts begin to accumulate at the soil surface. If moss is growing on the soil, it begins to die. Eventually the tree suffers too. Regular surface watering (make certain that water flows freely from the pot's drain holes before you finish) avoids this problem. Since city water is hard, I use rain water or water collected from a dehumidifier. When I run out of both, I used melted snow. My feeling is that bonsai can be grown indoors using water from the tap but it's more difficult.

Fertilizing is an ongoing process. Each time I water—unless I'm in a big hurry—the trees receive a light dose. How weak? I've been using products like Rapid-gro, Miracid, Wonder-gro and Peters 20-20-20 at a rate of one teaspoon in five gallons of water. Regular application makes up for the low concentration. Keep in mind, however, that even this very mild solution can kill dry plants. I've demonstrated this repeatedly. Water the dry plant with plain water first, then apply the fertilizer.

Incidentally, withhold fertilizer on the newly potted or repotted tree until it has definitely begun growing and is well on its way to recovery. The weak tree is more likely to be damaged than "saved" by fertilizer.

The ideal soil mix is an unresolved issue. Practicing bonsai enthusiasts certainly don't agree on a "best" recipe but do agree that soil is important. The goal is a growing medium with lots of internal "spaces" between the particles, spaces large enough to admit water read-ily and at the same time let excess moisture escape (allowing air to re-enter the soil). Of course, enough water should be retained to meet the plant's needs.

One way to obtain these internal spaces is to screen the mix ingredients before they are combined. Use a standard window screen and throw away anything that goes through the screen. Screening may be more important than the choice of ingredients.

Lacking a "best" recipe, one can do much worse than the time-honored horticultural blend of one-third coarse sand, one-third sphagnum peat and one-third "good garden soil" (all by volume). The preferred soil is a sandy loam. Avoid anything clayey. Packaged potting soils are quite variable in composition and quality so it may be best to use them only as a substitute for "good garden soil" in the 1:1:1 ratio mix, if they are used at all.

The combination I currently favor is one part grit ("starter" chicken grit from the feed and farm supply store) or sand, two parts sphagnum peat (available at garden centers) and three parts Turface (a commercial horticultural product consisting of baked clay particles).

Another consideration is inadequate light as a result of not keeping the bonsai close enough to the fluorescent unit. Light intensity diminishes very rapidly with increasing distance from its source. Accordingly, the uppermost parts of one's tallest bonsai should almost be brushing the tubes. My personal guideline is that no part of any tree should be further than 12 inches away. Since the light output of a fluorescent tube drops greatly with extended use, it's a good practice to replace the tubes once a year. Replace one tube in a fixture, then replace the second tube a few days later to avoid any damage to plants sensitive to the light increase.

Plant selection is the final issue. The first plants I brought indoors were some familiar outdoor kinds—such as firethorns, cotoneasters and boxwoods —which I thought would not require annual cold-dormancy to thrive.

Then I began receiving gifts of subtropical and tropical plants from friends. Sometimes I received a cutting, sometimes a small plant newly established in a plastic container; in most cases, the types of plants that would not survive winters outdoors in Michigan. I have enjoyed becoming acquainted with these "foreign" plants.

I have a Top Ten List for indoor bonsai—boxwoods (*Buxus* species), cotoneasters (*Cotoneaster* species), false heather (*Cuphea hyssopifolia*), figs (*Ficus* species), English ivy (*Hedera helix*), Greek myrtle (*Myrtus communis*), firethorn (*Pyracantha*

Juniperus procumbens 'Nana', three inches high, a 14-year-old cutting grown under lights for nine years.

coccinea), azaleas (*Rhododendron* species), Chinese sweetplum (*Sagaretia thea* [*S. theezans*]) and serissa (*Serissa foetida*)—but believe me experimentation is a great deal of the fun in bonsai growing. The more kinds you try, the more fun you'll have and ultimately, the more bonsai you'll have. 🌳

EDITOR'S NOTE: CULTURAL INFORMATION ON MR. WIKLE'S TOP TEN MAY BE FOUND THROUGH-OUT THIS HANDBOOK.

WHAT MAKES A GOOD BONSAI?

❶ The container is of a style, shape and color to complement the style of the tree.

❷ The surface roots, if any, make a gentle pattern radiating from the base of the trunk. No roots are crossed one over the other, nor are any exposed in an extreme or unnatural manner unless this is in keeping with the style of the tree.

❸ The trunk is positioned in the container in an aesthetically satisfying spot for its particular style. Approximately the first (bottom) third of the trunk is clearly visible, and the second third is partially visible. The trunk tapers from the earth to the tip of the tree. There are no abrupt or artificial changes.

❹ The main branches are gracefully arranged left, right and rear of the trunk. The distance between them is of equal or nearly equal proportion on all parts of the tree. None crosses another.

❺ The twigs which grow from the branches make delicate and precise patterns, all of about the same length. If there are training wires, they are applied neatly to both branch and twig. The wires are of a dull color so they do not disturb the overall effect more than necessary. A "finished" bonsai has no wires.

❻ There is no evidence of stubs left from pruning or marks from wires, weights or other props used in training.

(ADAPTED FROM **TROPICAL BONSAI**, AMERICAN BONSAI SOCIETY, 1967)

A   B

Two trees from nursery.
A shows more potential than B because of the more numerous, better spaced branches.

ILLUSTRATIONS BY JEANNE DERDERIAN

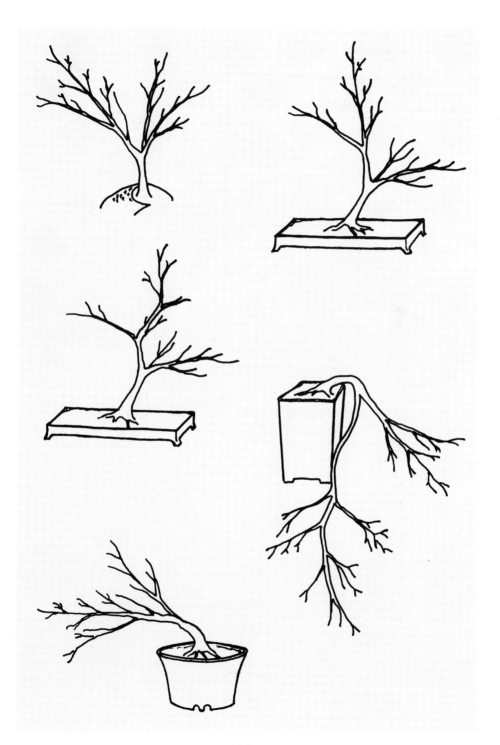

Small bamboo and *Juniperus chinensis* var. *sargentii* from BBG's bonsai collection.

GROWING WITH ARTIFICIAL LIGHT

MARGERY M. CRAIG

Many indoor bonsai can be grown successfully under fluorescent light. In fact, plants often do better there because their cultural needs may be more easily met in the controlled environment of the light garden than in the varied conditions of windowsills around the home. Ideally, the light garden provides a winter growing area for subtropicals that have summered outdoors in full sunlight, but for apartment dwellers who have limited outdoor space, the lights can be used year-round.

Plants to Grow

A great variety of plants will adapt to fluorescent light culture. *Serissa* grows and flowers continuously during the winter months, while gardenia (*G. jasminoides radicans*), Barbados cherry (*Malpighia glabra*) and dwarf pomegranate (*Punica granatum nana)* also bloom, but less frequently. Plants native to very warm climates like bougainvillea bloom under fluorescents in late January; pyracanthas and azaleas from milder parts of the temperate zone bloom in early spring.

Among the evergreen plants that grow throughout the winter are podocarpus,

MARGERY M. CRAIG, *Brookline, Massachusetts. A research biochemist by profession, she has devoted many years to the study of bonsai.*

various cypresses, Japanese box (*Buxus microphylla*), weeping fig (*Ficus benjamina*), creeping fig (*Ficus pumila*), English ivy (*Hedera helix*), olive (*Olea europaea*), junipers and false-cypresses (*Chamaecyparis*). Other types of plants will rest for two or three months during the winter, then start active growth in early spring. Examples of these are Natal plum (*Carissa grandiflora*), willow-leaved fig (*Ficus neriifolia regularis*), *Camellia sasanqua*, brush-cherry (*Syzygium paniculatum*; often referred to as, *Eugenia myrtifolia*), Chinese elm (*Ulmus parvifolia*), limeberry (*Triphasia trifolia*), lantana and hibiscus.

The light garden may also be used to start cuttings and seeds for indoor or outdoor bonsai. A clear plastic box containing a two-inch layer of coarse, screened perlite makes a convenient cutting box as well as a good place to store fingertip-size mame (miniature) bonsai during long weekends and vacations. The perlite, which is moistened, maintains a humid atmosphere but is sterile so that damping-off or other fungus growth is seldom a problem. If there are two or three small holes in the top and in the bottom of this box for air circulation and drainage, mame bonsai can be kept alive for a week or two without attention. As a cutting box, it rarely needs any care. After the initial watering when the cuttings are made, it should be checked monthly, but no additional water should be added unless the perlite is dry to the touch.

Design

The design of the light garden can be as varied and decorative as space and ingenuity allow. It has the great advantage of being flexible in size so that units may be added as desired. Commercial fluorescent units are available at garden centers for those wishing to design their own; most of the necessary materials are available at hardware or building-supply stores.

In choosing the length of fluorescent tubes, it is useful to remember that light intensity diminishes at the ends of any fluorescent tube. For this reason longer tubes are more efficient than shorter ones. The 48-inch tube is a convenient basic unit and the light garden may be made up of multiples of this readily available size.

A typical light system consists of two 48-inch tubes mounted in single strip fixtures spaced six inches apart on the underside of a shelf or stand 20 inches above a table or the shelf below. Two standard 11-by-22-inch waterproof plastic plant trays fit conveniently under each pair of lamps. To give varied distance to the lights for different size plants, one of the trays may be set up three to four inches on an overturned flat or on blocks of wood. The ballast from the fixtures may be removed and mounted separately to save space and reduce heat in the plant area. A reflector above the lights is not required if the shelf on which the tubes are mounted is painted with super-white flat paint.

Green plants cannot use all wavelengths of visible light. The red and blue ends of the spectrum promote plant growth and flowering, so fluorescent tubes which provide greater light intensity in these areas will produce more lush growth. I have found that Verilux Tru-Bloom tubes are very satisfactory, as are the Gro-Lux Wide Spectrum bulbs. The improved deluxe cool white and warm white tubes now also have increased output in the wavelengths plants can use. For maximum growth the tops of the plants should be no more than four to six inches from the lights. An appliance timer set for 14 to 15 hours per day can be used to turn on the lights automatically.

Growth Requirements

The humidity required for healthy plant growth should be provided by placing a half-inch layer of pebbles in the bottom of the plant trays and keeping this covered with water. To prevent the soil from taking up this moisture, the pots should rest on a support above the level of the water. A convenient support may be made from "egg crate" nylon fluorescent light diffusers available at building supply stores. Saw it to fit snugly into the tops of the plastic trays which taper and support the diffuser above the water level. If the temperature in the growing area ranges between 60-65° F this arrangement will provide sufficient humidity. Some plants require greater heat and humidity. This may be obtained by taping a 75° F heating cable in the bottom of the trays before the pebbles are added.

The cultural requirement for successful light gardening with indoor bonsai is to have good air circulation both in the air around the plants and in the soil. The first need can best be met with small fans designed to run at low speed. One fan hung in each 48-inch unit moves enough air to keep the plants healthy. On the second point, the soil mix must be carefully prepared and proper attention given to watering only as often as the soil dries. This will vary with the size of the pot. Very small bonsai require water every day. Most larger plants need water only once or twice a week.

TYPES OF FLUORESCENT TUBES

COOL WHITE & WARM WHITE
least expensive
average life – one year

AGRO LITE & WIDE SPECTRUM GROLUX
broad spectrum
expensive
longer life

VERILUX, VITALITE & G.E. CHROMA 50
broad sun spectrum
more expensive
much longer life – up to four years

**HIGH PRESSURE SODIUM,
LOW PRESSURE SODIUM
& SYLVANIA SUPER METAL ARC**
high intensity discharge
very costly

FOOT CANDLE READING WITH A PHOTO ELECTRIC EXPOSURE METER

Point the meter at a clean white paper at a distance of approximately 8 - 10 inches above the surface. Be certain not to cast a shadow from the light source on the paper. Set the meter at A.S.A. 10. With the exposure meter set for 1/100 of a second, the following are the foot candles:

METER READING	FOOT CANDLES
f 3.5	400 F.C.
f 4.0	500 F.C.
f 4.5	650 F.C.
f 5.0	800 F.C.
f 5.6	1000 F.C.
f 6.3	1300 F.C.
f 7.0	1600 F.C.
f 8.0	2000 F.C.
f 9.0	2400 F.C.
f 11.0	4000 F.C.
f 12.7	5200 F.C.
f 16	6400 F.C.

ARTIFICIAL LIGHT REQUIREMENTS FOR INDOOR BONSAI

JOCHEN PFISTERER

P lants used for indoor bonsai originate in warm climates. Plants of these regions have different care requirements when cultivated indoors.

• TROPICAL RAINFOREST

Weather in a tropical rainforest is not so humid as the name suggests. In the morning there is a hot and drying sun; the notorious tropical thunderstorm does not occur until afternoon. Young trees must grow in the shadow of their older neighbors.

With their adaptation to moist conditions and shallow light, plants of this cli-

mate are easily grown indoors. Their leaves accept dry air (central heating) because of a thick cuticle; artificial light of 700-1000 Lux* is sufficient as they are accustomed to half-light.

Ficus species and *Schefflera* are trees of this type; the tropical shrub *Polyscias* accepts dim light but it's leaves require high humidity.

• SUBTROPICAL FORESTS

These forests are not so dense—sunlight comes down to the floor. Each year there are two rainy (monsoon) periods and two dry spells. During dry periods many trees drop their leaves. But coastal sub-

JOCHEN PFISTERER, *owner of a bonsai nursery in Baden-Baden, Germany, is also a biologist, gardener, bonsai student and teacher, and author of many articles on bonsai as well as botany and ecology.*

*To convert Lux to foot candles, the unit of light measurement more commonly used in the U.S., multiply by 0.4. For information on how to use a photo electric light meter to measure light available in a particular spot, see the chart on page 23.

A closet equipped with shelves and fluorescent lights can be adapted
for growing bonsai.

tropical sites have no real dry periods and these plants need high air humidity all year. These include the well-known *Serissa, Ehretia* and *Sageretia* while *Ulmus parvifolia* accepts dry air better.

Subtropical plants need 2000-5000 Lux light and a higher air humidity (50% minimum).

• MEDITERRANEAN CLIMATE

Subtropical conditions prevail: Rainy periods in autumn and spring with a short dry spell in winter and a long dry and hot period in summer. Similiar climate is found in southern California, Chile, South Africa and the southwest coast of Australia. Typical plants are sclerophyllous shrubs and small trees like *Olea, Myrtus, Cistus* (Mediterranean), *Acacia, Callistemon* (Australia), *Phylica* (South Africa), *Lippia* (Chile), *Schinus* (California). These plants accept dry air, but need brilliant light—up to 10,000 Lux. In winter place plant in southern window or add additional electric light.

• SUBTROPICAL SEMIDESERTS:

Plants from these dry regions, succulent and sclerophyllous plants, need bright light—more than 10,000 Lux. Their weak roots do not accept a wet soil; they must be watered cautiously. Trees of this climate include the well-known *Adenium, Crassula* and *Portulacaria*.

Lamp & Light Color Recommendations

The best light for any plant is normal sunlight in an open-air location, if temperature is sufficient. The best artificial light is that which copies natural sunlight and this is found in lights with a light yellowish white color.

Common filament lamps emit such color light, but only for our eyes. This lamp type also emits infrared light at a high level and infrared radiation means heat.

Aquarium operators know about lamp problems. They use fluorescent lamps in two different light colors: WARM-LIGHT—a light yellow, and NORMAL-WHITE or DAY-LIGHT-WHITE—lights which contain all colors of the rainbow in equal concentration.

The pale-violet "PLANT LIGHT" recommended by the industry does not provide as good results as the light colors mentioned above. Leaves are green because of chlorophyll. This pigment absorbs the light colors red, orange and yellow, blue and violet. Green is reflected (this is why leaves look green to our eyes). The violet color of "plant lights" copies exactly these light colors which chlorophyll absorbs rather than reflects. A leaf also contains supplementary pigments— for example, carotene. This pigment is yellow and also absorbs green light. This is why light color lamps approximating sunlight give better results.

During the last 15 years I have grown dozens of species of indoor bonsai under electric light. Under "plant light" I had to move a bonsai within four months to the greenhouse for rejuvenation. But under one single 15W flurorescent lamp, Phillips "WARM DE LUXE" 14 hours a day, a *Ficus retusa* ssp. *crassifolia* has thrived for a minimum of five to six years without any problem.

If you have room for only one fluorescent lamp, use a "warm" light color. For two lamps, one "daylight" (or "normal white") and one "warm" is recommended. For three lamps, two "daylights" and one "warm", or one "normal white" and two "warm" lamps is recommended. Change fluorescent lamps every six months as after this time their power diminishes.

Spot Light

Because of their size, fluorescent lamps are only good for an oblong site such as a sideboard or niche.

To illuminate a single indoor bonsai placed on a desk, for example, a spotlight is more beautiful. Here a mercury arc-lamp is recommended. Select those sold especially for plants. The light of these lamps is very bright (1000 - 2000 Lux minimum), but sorry to say, all these lamps emit a slightly violet light and after some months the leaves of a tree, even when accustomed to dim light like *Ficus schlechteri*, will change to a pale grayish green, first sign that the light is not sufficient.

If an electric lamp is not bright enough, the leaves will eventually turn yellow and new branches will make long internodes. My advice is not to grow any plant under electric light of any violet color for longer than six months. Select another plant for this site and let the first one regenerate at a site with clear daylight.

How Long Light Has To Be On

A tropical day lasts 12 hours. So tropical plants are accustomed to having light for this amount of time. If the power of your lamp is not sufficient, turn it on for 14 hours a day but no longer as plants need a night as well. A timer will help for regular phases.

Additional Light for a Window with Dim Light

A northern window or one shaded by trees or a neighbor's house will not offer sufficient light for bonsai needing brilliant light. An additional lamp hanging overhead may help. On shorter winter days artificial light may prolong the plant's day time and the additional light from the top will encourage the leaves to turn not to the light outside, but rather to the light above, a very natural position.

SECURING BONSAI IN A POT

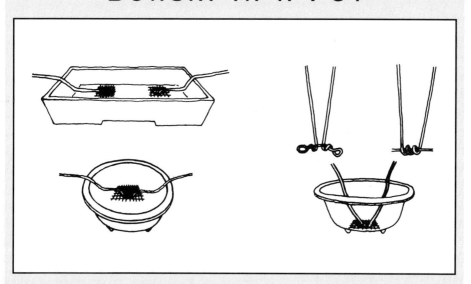

If the pot has only one drainage hole, twist wire like a candy wrapper or wrap it around a finishing nail so it will straddle the hole and be anchored to the underside of the pot.

DESIRABLE QUALITIES OF INDOOR BONSAI

SIGMUND I. DREILINGER

Many different species of trees, shrubs and vines can be trained as indoor bonsai. There are five qualities that are important:

❶ The ability to grow under reduced light.

❷ Lack of the need for a cool or dormant period. The above factors suggest understory trees—subtropical and tropical species.

The other three qualities of slightly less importance are:

❸ Attractive bark—such as *Malpighia*, Cork Bark Chinese elms.

❹ Flowers and fruit—such as *Carissa*, *Serissa*, *Malpighia*, pomegranate.

❺ Small or reducible leaf size—such as Chinese elm, Kingsville box, sea grape.

Temperate zone bonsai can be grown indoors successfully, but only when a

chilling or dormant period can be supplied. I have succeeded in growing *J. chinensis* var. *sargentii*, *J. squamata* 'Prostrata' and *J. procumbens* 'Nana' indoors. This was done by keeping a window open, so that the temperature was lowered to 45 or 50 degrees F at night. Daytime temperatures remained at 70 to 75 degrees F.

Small Flowers & Fruit

Since these do not reduce in size readily, if at all, it is best to work with trees that have small fruit and small flowers.

Many trees are photoperiodic. Flowering is affected by the length of the day or night. Greater or lesser fruiting can sometimes be enhanced by the use of hormones or chemicals such as "Tomato Set."

Small Leaf Size

Small leaf size is more in proportion with the size of the bonsai. Defoliation can further reduce leaf size. Sea grape leaves have been reduced from seven or eight inches to one inch by this method.

NOTE: THE LIST OF TREES IN THIS HANDBOOK WILL INDICATE WHICH ONES CAN ADAPT MOST EASILY TO INDOOR CONDITIONS.

SIGMUND I. DREILINGER, *President and founding member of the Bonsai Society of Greater New York, editor of* **The Bonsai Bulletin***, past president of* **Bonsai Clubs International***, author and teacher, student to bonsai masters, is Guest Editor of this Handbook.*

Pomegranate, part of the Naka collection, has small leaves and interesting trunk, flowers and fruit—all desirable bonsai qualities.

PHOTO BY SIGMUND DREILINGER

FERTILIZERS

SIGMUND I. DREILINGER

Many bonsai growers no longer use soil as a planting medium, but rather a soilless mix. This necessitates regular use of fertilizers, either organic or inorganic.

THE ADVANTAGES OF INORGANIC FERTILIZERS:

❶ They are water-soluble and readily absorbed by the roots.

❷ Use of different products will usually supply some trace elements in the impurities that are present in all chemical fertilizers.

❸ Trace elements can be added as needed.

THE DISADVANTAGES OF INORGANIC FERTILIZERS:

❶ If used in too great a strength they will burn (injure) the root system.

❷ Daily watering will rapidly leach these soluble salts out of the planting media.

THE ADVANTAGES OF ORGANIC FERTILIZERS:

❶ There is a slow, small, steady release of fertilizer elements with each watering.

❷ No danger of burning the roots.

❸ There are small amounts of trace elements in organic material.

❹ They are long lasting.

THE DISADVANTAGES OF ORGANIC FERTILIZERS:

❶ They have an offensive odor as they decompose.

❷ They attract insects.

❸ They burn the moss.

If you use inorganic fertilizers, vary the manufacturer. Use coated or slow- release types. Use regular chemical fertilizers at less than the recommended strength. When bonsai are in active growth feed every two to three weeks. Reduce the frequency when growth slows; once every six to eight weeks is often enough. To reduce the danger of salt buildup, every four to five weeks water for one week with plain water.

Chemical fertilizers usually have a neutral pH of about 7.0. Acid fertilizers have a pH of 6.0 to 6.5. Read the labels of the package to make certain which fertilizers to use with which bonsai.

Nitrogen, phosphorus and potash percentages of each package are expressed in numbers such as 5-10-5. There are many different mixes available for different purposes. For more growth use a fertilizer with more nitrogen such as 10-6-4. For flowers and fruiting more phosphorus is better, such as 10-30-20.

After transplanting or root pruning do not fertilize for three to four weeks. There are vitamin solutions that are specifically made for use in reestablishing bonsai. One such is Transplantone. Others contain Vitamins B-1. Another chemical fertilizer is BR 76, with a 9-59-8 composition. It is good for blooming and fruit growth and root growth, but use it at one-half strength.

The pH of your water may cause chemicals to become locked up and unavailable. For most bonsai, a pH of 6.5-7.0 is best. For acid-loving bonsai, such as azaleas, 6.0-6.5 is needed. Inexpensive pH meters are available, or pH papers, or pH test kits. Pool supply centers are a good source for these materials. You can also phone your city water supply source and ask about your pH. In South Florida, Texas and other areas, the water is so alkaline it may reach pH 8.0-8.5.

PLANTING MEDIA

Sigmund I. Dreilinger

The needs of the commercial nurseryman have led to many changes in planting mixes in the last thirty years.

Soil, with its many insects and diseases, requires sterilizing treatment and some fertilization. To alleviate such time-consuming problems, soilless mixes were developed. Today—if it's used at all—sterilized soil is a mere 10 to 15 percent of the total composition of planting mixes. Other components include:

PEAT MOSS: slightly acid and retains water.

COARSE BUILDERS SAND: hard, neutral, does not absorb water, holds water on surface.

PERLITE: heat-treated mica-type rock, (usually in flakes), lightens mix, neutral, absorbs water.

VERMICULITE: heat-treated, mica-type rock, usually in flakes. Over time it tends to compact.

TURFACE, TERRAGREEN, HAYDITE AND OTHER FIRED CLAY PARTICLES: neutral, porous, absorb water, do not break down.

CHICKEN GRIT, TURKEY GRIT: basically coarse sand, hard, does not break down: sometimes contains other matter such as ground shell.

DECOMPOSED GRANITE: used a great deal in California; hand sieve to get rid of small particles; hard.

PINE BARK: use small size; porous, holds water, breaks down and decomposes within a year. Do not use more than 10 to 15 percent.

FIR BARK: use small size; does not hold as much water as pine bark; will last longer than pine bark before decomposing.

CHATTACHOOCHIE: small gravel or pebbles of over 1/8" to 3/16" diameter, hard, non-porous, used a great deal in Florida.

CRUSHED LAVA ROCK: neutral, porous, absorbs water; sieve to get rid of fine dust.

There are as many variations of planting mixes as there are bonsai growers. The most important factor is not water retention but airiness and ease of water penetration and drainage. Fertility is not a factor in any mix as it is controlled by frequency of use and variation in the components of the commercial fertilizers used by the bonsai grower.

Experimentation will determine which is the best mix for your bonsai. I use 1 part chattachoochie, 1 part Texas grit, 1 part crushed lava rock, 1 part pine bark (or fir bark). For acid-loving trees add 2 parts peat moss. When bonsai are in active growth I fertilize at half strength once a week for three consecutive weeks. The fourth week I use water with no additives to prevent salt buildup.

WATERING

SIGMUND I. DREILINGER

Japanese bonsai masters claim that it takes three to four years to learn the intricacies of watering.

Apprentices to the masters begin as "gophers." They are occupied with fetching, carrying and janitorial duties. It is in the second year that they first really commence to learn the how and when of watering.

In the United States there isn't an apprentice system but a teacher/pupil relationship. Without the daily interaction as provided by the Japanese apprentice system, there is an incorrect tendency among Americans to discuss watering as a matter determined by the clock.

But frequency of watering bonsai should take into account many factors. Some of these are:

❶ **HUMIDITY.** Centrally heated or air conditioned homes will sometimes have the relative humidity of a desert. Locating the container on a tray of gravel partially filled with water, or misting the bonsai 4-5 times a day, or the use of a vaporizer, will increase the humidity. Bonsai media have the tendency to dry out rapidly and will require more attention.

❷ **TEMPERATURE.** Higher temperatures will cause the bonsai to use up the available water supply and require replenishment.

❸ **PLANTING MEDIA.** Media that contain a greater percentage of larger particles of hard non-porous material such as chattachoochie, gravel or decomposed granite will retain less water. This increases the frequency of watering.

❹ **FOLIAGE MASS.** Other factors being equal, bonsai that have a greater foliage mass will transpire more, dry out sooner and require more water.

❺ **SIZE AND GLAZE OF CONTAINERS.** Small containers hold less media and consequently need watering more frequently. Containers that are glazed will dry out more slowly and will need less watering.

❻ **GROWTH AND DORMANCY.** Bonsai require more water during their growth period; when dormant, water less.

To some extent all these factors affect bonsai water requirements. This is why watering can vary from twice daily to as infrequently as every second day.

The easiest way to keep track of the condition of your bonsai medium is to insert your finger into it and feel whether your medium is damp. Lifting the container with your bonsai in it will also determine whether it needs watering. When dry, it will be lighter than when wet. An inexpensive water meter may be purchased from a nursery and I find its use an accurate way to determine the condition of the planting medium.

To create a beautiful bonsai, nothing takes the place of close and careful observation on a daily basis. 🪴

Serissa foetida in flower.

PHOTO BY ELVIN MCDONALD

PESTS & PROBLEMS

Before adding a new plant to your bonsai collection, keep it in quarantine for two to three weeks and examine it daily for pests.

If you summer your bonsai outdoors, combine acclimatization with preventive treatment. Place your bonsai in a shaded place for two weeks before bringing indoors. Check them very carefully for pests, especially the axils, undersides of the leaves, and the crotches of the branches.

The use of the correct insecticide is

HOUSEPLANT PROBLEMS

DAPHNE S. DRURY

LEAVES WILTED	Too much or too little water. Too small or too large a pot.
LEAVES DROPPING OFF	Too much water.
BOTTOM LEAVES ARE YELLOW	A few are to be expected. If there are many, the plant may be potbound or may be going into a resting period.
LEAVES ARE PALE GREEN OR YELLOWISH	Too little or too much light. Lack of fertilizer, particularly nitrogen.
PLANT THIN AND ETIOLATED	Too little light.
GROWTH STUNTED	No drainage hole. Too heavy a soil mixture.
LEAVES HAVE DRY TIPS	Humidity too low or soil too wet.
LEAVES ROLLING UP	Drafts, especially from an air-conditioner or ill-fitting windows.
YELLOW OR BROWN SPOTS ON LEAVES	Caused by sun burning the leaves.
SUCCESSIVE LEAVES GROWING SMALLER	Potbound or insufficient fertilizer.
PLANT ROTTING AT NECK OR CROWN	Too much water. Badly drained soil. Too cold.
ALL LEAVES DROP OFF SUDDENLY	Cold air. Gas injury. Lack of water.

PESTS

NAMES	SIZE	COLORS	STAGES	PESTICIDES*	COMMENT
APHID	$1/16$"-$1/8$"	Red, green, pink, brown	Egg 3-8 nymph stages Adult	Dormant oils Metasystox Malathion Pyrethrum	2-3 applications 3-5 days apart
SPIDER MITE	$1/32$" or smaller	Red, Brown, Black	Egg Larva Chrysalis Feeding stage Chrysalis adult	Dormant Oils Kelthane Metasystox Orthene	3 Applications 7 days apart. Difficult to see. Put paper under branches & tap the branches. Look for moving specks.
SCALE	$1/16$"-$1/8$" Oval or round bump	Brown, black gray, white	Egg Nymph Crawler Adult	Dormant Oils Malathion Orthene Metasystox	Crawler stage – easiest to control. Spray 3 times 7 days apart. Touch adult scales with Q-tip dipped in alcohol.
MEALY BUG	$1/16$"-$1/8$"	White cotton-like fluffs	Egg Adult	Malathion Orthene	Easily controlled. Spray twice 1 week apart. Check under leaves, axils or branch crotches.
WHITE FLY	$1/16$"-$1/8$"	White	Egg Crawler Feeding stage Winged adult	Resmethrin Orthene Malathion	Spray 3 or 4 times at 5 day intervals, more if needed. Yellow sticky boards will catch adults.

* Use pesticides only as a last resort. Try hand-picking and insecticidal soaps first.

imperative if pests are present. Check to ascertain that those used are not harmful to your bonsai. Read the labels carefully, and use only at the proper dilution.

When you move your bonsai indoors, examine them daily. If there is a pest problem, try Pyrethrum or Resmethrin or some of the Safer agricultural soaps. These are relatively harmless to humans. The use of contact or systemic poisons indoors is potentially harmful. If you must use them, spray in the garage or outdoors at midday if the temperature is above 40 degrees F. Use gloves and a face mask and wash off thoroughly and immediately (including your clothes). Systemics can be absorbed through the skin.

Do not use Malathion on *Ficus, Crassula, Podocarpus* or buttonwood. Do not use Dimethoate on *Ficus* or *Podocarpus.*

EVEN BONSAI GROWERS NEED A VACATION

EDMOND O. MOULIN

ecause bonsai are grown in small containers, they require more frequent watering than many houseplants do, and the grower must be ever alert to their needs. But what happens when you want a vacation? Try the "bonsai-sitter bag." It's also an aid for acclimating plants newly purchased from a greenhouse or to nurse recently root-pruned bonsai through a critical period. Here's how it works:

Use a clear plastic bag large enough to envelop the plant and container. Care

is essentially the same as for a terrarium. Water the soil well and let it drain thoroughly before placing container in the bag. Inflate the bag to create a bubble or form wire into hoops and insert the ends into the container to make a support frame for the plastic so that the foliage does not touch the plastic.

Place the enveloped plant where it gets good light but no direct sun rays.

The bag may have to be opened occasionally to allow excess moisture to evaporate. This works well for about two weeks without adding water. Open the bag gradually over several days to acclimate the plant to drier surroundings.

EDMOND O. MOULIN, *Director of Horticulture at the Brooklyn Botanic Garden, has written articles and lectured on bonsai.*

Serissa foetida variegata. Photo by Christine M. Douglas

GROWING FROM SEED & CUTTINGS

KATH WILLIAMS

A great deal of pleasure in growing bonsai indoors comes from the opportunities it provides to grow the more exotic plants. Here in Great Britain, as in many other parts of the world, bonsai enthusiasts have a wide selection now available to them from tropical, subtropical and other climates where bonsai originate.

Seeds

The pleasure that can be obtained from growing from seed enhances the benefits of growing indoor bonsai. A real advantage in growing subtropical plants from seed is that they germinate more readily and grow more quickly than our native, hardy specimens. The result is that you can have an attractive little bonsai quickly.

In stock at our local supermarket is a considerable variety of tropical and subtropical fruit. Some which are very straightforward to germinate and similar

KATH WILLIAMS, *a bonsai enthusiast living in Great Britain, has written many articles for bonsai society journals and is a lecturer and teacher as well.*

in character are the citrus fruits: orange, lemon, lime and grapefruit. They make very pleasant evergreen trees with dark shiny leaves. They germinate easily and show a rapid rate of growth to begin with, but do not be surprised if they slow down after their initial spurt. They soon get going again and proceed to grow slowly and steadily.

Another easy seed to germinate is the pomegranate. If you buy one fruit and plant a third of the seeds, you will still have more trees than you can cope with. Be careful not to overwater the young seedlings or they will rot and collapse.

The more unusual fruits one can try are fresh olives, mangoes, dates, figs and litchis. All of these will germinate provided you wash the seed and allow them to dry for a few days in a well-ventilated area. Soak them overnight before planting in a good seed compost, then set in a warm place. Use either a propagating tray, or failing that, the airing cupboard —this is ideal as long as you watch for the first shoots to appear, then place them in a light area. If left too long in a dark cupboard, etiolation results, producing the inevitable long, leggy seedling of little strength or attraction.

Once they've germinated, plant in individual pots in a well-drained mix comprising of 50% peat, 25% loam and 25% grit. With a fair measure of persistence and some luck, your seedlings should soon be reaching the stage when you may start training them. Pinch out and wire exactly as you would outdoor trees. Remember, however, that many of these indoor bonsai are evergreen and do not have a surge of sap in the growing season. Unlike native deciduous trees that become less brittle when sap flows, many indoor specimens remain brittle all year. They snap off very easily—therefore try to wire branches while they are still green and pliable.

Fig 1

Cuttings

One can start bonsai from cuttings or even retrain houseplants. Taking indoor bonsai cuttings is a year-round exercise as they grow year-round.

For those plants that take a rest period, wait until the new growth commences before taking cuttings. If your house is centrally heated to a temperature of 70° F. or more, and you maintain a good level of humidity around your indoor specimens, this may be your trees "warmest" season.

Take cuttings of new wood about four inches in length, dip into hormone powder, place in either perlite or sand (not too fine), moisten, cover with a plastic bag and place in a constantly warm temperature. Roots on most varieties should appear in four to six weeks.

Varieties from which cuttings can readily be taken include *Ficus* (weeping fig), *Crassula* (jade tree), *Schefflera* (umbrella plant) and *Dracaena*. Once rooted, treat like any other young stock.

Fig 2

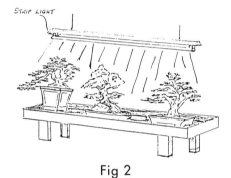

Fig 2

RESTING PERIODS, LIGHT EFFECTS AND INDOOR BONSAI

STEPHEN K-M. TIM

The days shorten and the air becomes tinged with a perceptible chill. Autumn has arrived. Our thoughts turn to the change in season in the garden and the protection of the plants over the winter months ahead.

Where indoor bonsai are concerned, the pros and cons of providing a resting period are not at all clear. In discussions with Edmond Moulin, Director of Horticulture at the Brooklyn Botanic Garden, who has taught bonsai over many years, he has expressed some concern over the confusion between the terms *dormancy* and *resting period*. Some bonsai growers use the terms interchangeably. Mr. Moulin and I consider them distinct and will attempt to differentiate between the two.

STEPHEN K-M. TIM *is Vice President of Science and Publications at Brooklyn Botanic Garden.*

Azaleas need a rest period of four to six weeks.

PHOTO BY ELVIN MCDONALD

Resting or Dormant?

Hardy bonsai, including the beeches, maples, junipers, firs, spruces and pines are, in nature, prevented from growing during the freezing to near-freezing winters. They undergo full dormancy and in this way survive the vicissitudes of this frigid season. Non-hardy bonsai, those being discussed here for indoor growing, are drawn from a great diversity of habitats, from the subtropics to the temperate regions of the world. Ones from the latter region enjoy cool but mostly frost-free winters with temperatures dropping to a minimum of 40° to 45° F (4° to 10° C). Providing a period of lower but above freezing temperatures over winter will be called a resting period in contrast to a dormant period where freezing or near-freezing temperatures are implied.

The need to fulfill the dormancy requirements of hardy bonsai is well recognized; without a period of exposure to cold, usually near freezing, these may defoliate, show shoot die-back and decline in general health. Providing a resting period for indoor bonsai has been less strongly advocated. This is most likely due to their responding less dramatically when a regimen of a resting period is not strictly observed. The exception is with plants that respond to a resting period by rewarding us with a dramatic spring flowering. In the case of Kurume azaleas, *Rhododendron obtusum*, a resting period of four weeks at 40° F (4° C) is necessary to assure a flush of flowers in spring, and the common olive, *Olea europaea*, fails to produce flower buds if denied a cool spell the previous autumn or early winter.

Even for plants native to the warm tropics, a uniform rate of growth throughout the average year seems not the rule. They appear to have a built-in or endogenous growth rhythm that, while not necessarily related to environmental necessity, appears beneficial to the plant if adhered to. The same applies to indoor bonsai. Mr. Moulin and I side with providing a rest period for all such bonsai, particularly ones from a warm temperate habitat. We may be hard-pressed to define the precise benefits to the plant but we feel the trees will respond by living healthier and longer lives.

Light Effects on Resting Periods

Light plays a part in the initiation of dormancy. In autumn, the shortening days trigger a reaction among temperate plants, causing growth to slow down and the leaves, in deciduous plants, to fall. This is all part of the process of acclimation, when hardiness becomes established enabling plants to survive the winter. In as much as dormancy may assure the plant's sustained vigor the following year, the length of day (length of the light period) to which a plant is exposed may influence its flowering. This is the Photoperiodic Effect.

Plants are classified as long- or short-day types, requiring more or less than 13 or 14 hours of light per 24 hours. By the correct priming of plants as their flower buds are being formed, their optimum in flowering is assured the following year. Day-neutral plants are independent of the length of day. Among these are the pomegranate and *Serissa* that appear to bloom irrespective of the length of the day/night cycle. Day length also influences autumn leaf coloration. Especially where red pigments are concerned, as in some maples, shortening days in late summer and early autumn, combined with bright days and cool nights, are necessary to bring out the most vivid of autumn leaf tints.

Preparing Plants for the Resting Period

When outside temperatures begin to dip

toward the 55° to 60° F (13° to 15° C) mark in the autumn, indoor bonsai need to be prepared for winter. If the plants have spent the summer outdoors and are to be brought indoors to a frost-free spot, the period of transition or preparation is a vital time. Avoid suddenly transferring them indoors. As summer comes to an end, move the plants to a shadier part of the garden for about two weeks. This primes them for living in lower light. Then bring them indoors at least two weeks before the heating system is to be turned on, enabling them to acclimate to the change in growing conditions. This slow transition is less important when the move is into a temperature and humidity controlled greenhouse but the growing conditions in the average home are extremely dry. The desiccating effect of the desertlike atmosphere is decidedly stressful on the plants, particularly if they are not adequately prepared for the change. The standard method of increasing the humidity around the immediate vicinity of the plants is to stand them on top of pebbles in a water-filled tray. This helps but a cool vapor humidifier is the only efficient way to raise the humidity indoors.

Plants and Their Requirements

Because of the wide range of habitats from which the more popular indoor bonsai are drawn, it is impossible to standardize their winter requirements. The grower is urged to use good sense in determining the conditions most suited to the particular plants being grown. The following is an attempt to aid in the process. Four categories are distinguished by their requirements for a resting period. The species have been drawn from the selection of plants most commonly grown indoors as bonsai. Other plants not listed should be matched with the appropriate categories.

CATEGORY I:
Tropical Plants

Includes:

Brassaia actinophylla :	Australian umbrella tree
Conocarpus erectus :	Buttonbush
Dizygotheca elegantissima :	False aralia
Ficus benjamina :	Benjamin fig
Polyscias fruticosa :	Ming aralia

Plants in this category are least dependent upon a defined rest period. Plant growth seems sustained at an even level throughout the year. Providing a resting period is optional but, as with all plants, at least a drop of about 10° F (12° C) at night is of benefit to the plants.

CATEGORY II:
Warm- to Mild-Temperate Plants

Includes:

Carissa grandiflora :	Natal plum
Crassula argentea :	Jade plant
Cuphea hyssopifolia :	False heather
Cycas revoluta :	Sago palm
Fortunella hindsii:	Hong Kong wild kumquat
Leptospermum scoparius :	Tea tree
Malpighia glabra :	Barbados cherry
Myrtus communis :	Common myrtle
Olea europaea :	Common olive
Podocarpus macrophylla :	Japanese yew
Punica granatum :	Pomegranate
Serissa foetida :	Serissa
Syzygium paniculatum :	Brush cherry

In their native habitats, these plants enjoy cool winters that are free from frost. A rest period of six to eight weeks with night temperatures of 50° to 55° F (10° to 13° C) and day temperatures no more than 10° F (12° C) higher during winter is optimal. In the confines of an apartment, finding a place with such a sustained temperature may be difficult. Check on temperatures near the floor below a window or move the plants to an unheated room. At the temperatures advocated, very little new growth will be made by these plants and they will need much less light than normal. But avoid direct, unfiltered sunlight as the sudden warmth may force plants into growth, making them vulnerable to subsequent drops in temperature. Keep plants moist but do not overwater. Occasionally mist the trunk and branches.

CATEGORY III:
Mild-Temperate to Temperate Plants

Includes:

Buxus microphylla :	Boxwood
Camellia sasanqua :	Sasanqua camellia
Chamaecyparis pisifera :	Sawara false cypress
Cupressus arizonica	Arizona cypress
Cupressus macrocarpa :	Monterey cypress
Ficus neriifolia :	Willow-leaved fig
Hedera helix :	English ivy
Ilex crenata :	Japanese holly
Ilex vomitoria :	Yaupon
Lagerstroemia indica :	Crape myrtle
Pinus halepensis :	Aleppo pine
Rhododendron indicum :	Indica and Satsuki azaleas
Rhododendron obtusum :	Kurume azalea
Sequoia sempervirens :	California redwood
Taxodium distichum :	Bald cypress

These plants are accustomed to lower temperatures than those in Category II and should be held at 40° to 45° F (4° to 7° C) for at least six to eight weeks. Growth will have almost ceased at these temperatures but leafy plants will be better kept where they receive filtered light to prevent leaf yellowing or leaf drop. Direct sunlight must be avoided. A garage, basement, unheated greenhouse or well-insulated cold frame will provide correct conditions. To insure that temperatures at the root ball remain even, pots may be covered with a mulch of sawdust, leaves, bark or coarse peat.

CATEGORY IV:
Temperate to Cool-Temperate Plants

Includes:

Juniperus procumbens 'Nana':	Juniper
Juniperus squamata 'Prostrata' :	Juniper
Pinus thunbergiana :	Japanese black pine
Ulmus parvifolia :	Catlin elm

While these plants can be treated as for Category III, they have been separated out because they can tolerate freezing temperatures during dormancy provided they are well protected. However, they are not ideal for indoor culture, especially for the inexperienced. Enthusiasts are too often tempted to try these plants because their form and compact leaf growth give the appearance of an "instant" dwarfed tree.

For bonsai growers intent on trying plants that are closer to being hardy bonsai rather than indoor types, and who do not have access to cool garages, basements or cold frames, the standard advice has been to overwinter them in a box on the fire escape of the average apartment building. If this is done, the pots must be mulched over and the open end of the box made to face the wall. This prevents desiccating winds from drying out the top growth and lessens the chances of cycles of freezing and thawing that are fatal to the plants. However, I should caution that the practice of placing any object on a fire escape is most likely illegal.

Chamaecyparis pisifera 'Squarrosa' needs a cool period
to grow successfully.

An alternative to the above is to refrigerate the plants for the required dormant period by housing them in that part of the refrigerator that holds a temperature of 40° to 45° F (4° to 7° C). Since the atmosphere here is dry, the plants should be placed in large inflated plastic bags on a layer of damp peat to provide moisture around the plants.

A word of caution here where Category IV-type plants have been subjected to full dormancy. Plants are better kept on the cooler side for an extended period of time, the reason being that if they are exposed to warm room conditions after their dormancy requirements have been met, the new growth that would result might coincide with the inadequate light intensity of January and/or February and etiolated growth may be produced.

GUIDE TO CERAMIC BONSAI CONTAINERS

PHIL TACKTILL

B onsai containers are a subject seldom touched upon by authors of bonsai books. Because the selection of a container is so important in terms of both aesthetics and cost, we should have some understanding of the vast variety of containers—their styles, colors, shapes and prices. The following illustrations and charts should be helpful references for selecting a container that is appropriate for the bonsai and for judging its quality and value.

First, some general information that is usually true:

- Japanese ceramic containers are fired at 2000° F—a higher temperature than used in other countries.
- Poured containers are the least expensive.
- Press-molded containers are medium to high priced.
- Thrown containers are medium to high priced.
- Hand-formed containers are in the high price range.
- Antique containers are the most expensive and most difficult to identify.
- Each additional operation in manufacture will add to cost.

Illustration of Containers

If you note the illustrations, it becomes obvious that each addition to the basic form of a container adds to the cost. The finer the detail, the more the cost. Chart A shows three views: side, top, bottom. The right side is the simplest, and the left side is the more ornate. The additional details added to the basic containers illustrated are as follows: #1 a cloud leg, #2 a bottom rim, #3 a window panel recess, #4 drawings and designs on the body (a right side drawing is shown), #5 an upper lip, #6 a notch in the corner of container with a rim. Add a glaze and you have just run the price of a simple container way up.

Some of these details can be molded in a poured container but are usually not in sharp detail. Number 7 shows construction of a container design that can be poured or press-molded. Its design is such that it can easily be withdrawn from the mold. Note the top of the design indents and the bottom protrudes so it can be slid from the mold.

Irregular forms, if used properly,

ILLUSTRATIONS BY PHIL TACKTILL

Chart A

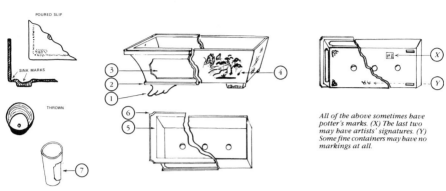

All of the above sometimes have potter's marks. (X) The last two may have artists' signatures. (Y) Some fine containers may have no markings at all.

			DISTINGUISHING FEATURES				
TYPE OF CONTAINER	SINK MARKS	ANY SHAPE	NORMALLY ROUND	THIN & LIGHT	THICK & HEAVY	GLAZED	UNGLAZED
poured slip mold	Yes	Yes		Yes		Yes	Yes
pressmold		Yes			Yes	Yes	Yes
thrown			Yes	Yes	Yes	Yes	Yes
hand formed	Yes	Yes		Yes	Yes	Yes	Yes

should be an asset, and in Japan they are prized possessions. They are what makes one container different from all that are produced. Much can be said for individuality of containers.

Matching Tree Style and Container Forms

Generally speaking, straight lines on a container go best with a straight (upright) tree or trees. Curved lines go best with informal trees. Chart C shows suggested combinations of tree styles with container styles.

The black dot in the illustration on Chart C indicates the recommended placement of the tree in the container. A more ornate bonsai container requires a more powerful or dramatic tree. The container should harmonize and complement the tree. The color of the container is most important. With ever-

A collection of bonsai shows the many types of containers that can be used.

greens, red (terracotta), brown and green body colors are recommended. with fruiting and flowering bonsai and trees with bright fall colors, choose a container with a color that complements the most colorful phase of the tree. (See Chart **B**.)

When using bright colored contain-ers, consider, for example, the effect of looking at a blue glazed container, 4" or 6" deep by 20" long. This long panel of color would require a tree that can con-trast with such a large colored area. On the other hand, the area of a *shohin* bon-sai container, 2" x 2", that is bright red would not be offensive, since the colored

area is so small.

Trees should be displayed at their time of maximum color.

There are many rules on how long and wide a container should be. I find one's eye to be a good judge and there is a wide latitude in selection. However, the depth of the container as it pertains to the thickness at the base of the trunk is of major importance. The formula that I use is the depth of the container should be from 1/2 to 2-1/2 times the thickness of the base of the trunk.

An obvious exception would be the semicascade and cascade trees where the greater depth of the container is an important counter-balance to the mass of foliage outside the container.

Some other points:

- If you display a square container with its point forward, you increase the length of the container.
- Vive la difference. I, like many, welcome a container that differs. What Americans refer to as an irregular container (a warped container for example) I look at as one-of-a-kind.

That makes my container unique.

- Containers should be stored outside and aged like the bonsai so that they will develop a patina and softened colors.

REPRINTED WITH PERMISSION FROM *BONSAI BULLETIN*, VOL. 26, NO. 1, 1989. FOR MEMBERSHIP INFORMATION, CONTACT BONSAI SOCIETY OF GREATER NEW YORK, MEMBERSHIP SECRETARY, P.O.BOX 565 GLEN OAK, NY 11004.

CHART B

Color of Container	Color of Fruit, Flower, & Foliage							
	White	Violet	Pink	Red	Yellow	Blue	Orange	Green
Red	•				•	•		•
Blue	•	•	•	•	•		•	
Green	•				•		•	
Yellow		•	•			•	•	•
Violet	•	•	•				•	
Brown					•		•	•
White	•		•	•	•	•	•	•
Orange						•		•
Black	•	•	•	•	•	•	•	•

50

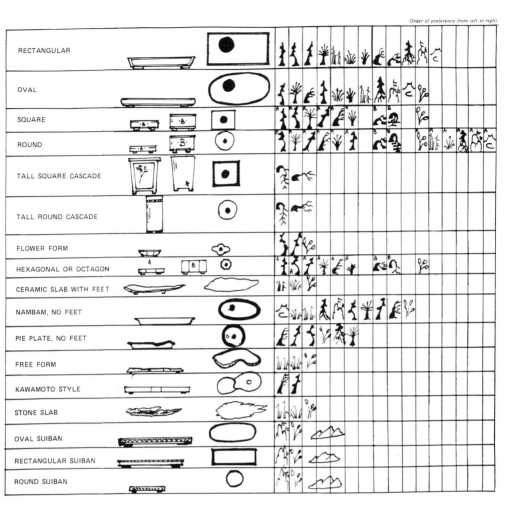

CHART C

		SINGLE TREE STYLE
		Formal Upright
		Informal Upright
		Slanting Style
		Wind Swept
		Semi Cascade
		Cascade
		Broom Style
		Root Over Rock

		GROUP PLANTINGS
		Formal Forest
		Informal Forest
		Clump
		Rock Planting
		Saikei
		Suiseki
		Accessory Plants Such as Grass, Bamboo

51

HOW TO INCREASE HUMIDITY FOR SUBTROPICAL BONSAI

JOCHEN PFISTERER

Bonsai-terrarium made of glass plates. The small strip at the bottom in front makes a flat tray, that can be filled with gravel to increase air humidity. Plants: *Ficus benjamina* 'Starlight', *Ulmus parvifolia*.

any subtropical and some tropical trees, grown as indoor bonsai, require high humidity because of their tender leaves, a result of thin cuticles. (This wax layer, when thick enough, protects leaves from drying out.)

These thin-cuticled trees are not easily grown in centrally heated apartments, especially those plants originating in subtropical coastal sites. Some examples are California cypress (*Cupressus macrocarpa*), plants from southern China and Japan such as *Ehretia microphylla*, *Sageretia thea*, *Serissa foetida* and *Ulmus parvifolia*. Shrubs from tropical rainforests—such as *Polyscias* species—are also unaccustomed to dry air.

With the exception of *Polyscias*, the plants cited require bright daylight during wintertime, in a window facing the south or west. During summertime a window to the east or west is recommended, although the optimum in summer would be a partly shaded site in the open air. *Polyscias* prefers a not-too-bright light. Electric light or a place in a window facing north, east or west is good.

In wintertime, air in a centrally heated room is rather dry. This is because a cold gas is not able to contain the same high level of relative humidity as a warmer gas. In wintertime the air outdoors has a normal humidity of about 60% or more. When this frosty air is heated indoors up to 20° C, its relative humidity sinks remarkably.

In a dry atmosphere only plants with thick cuticles will survive without prob-

A "tokonoma" in my dining-room. It is illuminated with one fluorescent lamp of only 15 W, sufficient for tropical *Ficus* species.

lems: the tropical *Ficus* species; the plants from subtropical semideserts such as *Brachychiton, Portulacaria* or *Sarcocaulon*; and the sclerophyllous shrubs such as *Laurus, Olea* and *Myrtus* from Mediterranean regions.

Fortunately, there are some easy methods to increase humidity around semitropical bonsai...up to 60 to 80%:

❶ INCREASED WATERING

This allows the roots to transport more water to the leaves. Of course there is a certain risk. Roots in a wet soil, because of anaerobic conditions, can rot. They do not receive enough oxygen.

To avoid this problem, you should use a soil mixture able to maintain good ventilation even when wet:

1/4 GRAVEL, 1-6 mm diameter (1/25 - 1/6 in.) of lava, pumice or broken bricks. Do not use gravel containing lime. Fine gravel guarantees a better ventilation of the soil than does sand.

1/4 PEAT, (short fibers are acceptable)

1/4 COMPOST, sifted with a screen, meshes up to 1/3 - 1/2 in.

1/4 CLAY, in small grains

This soil mixture even allows permanent watering of a bonsai by wicks from a tray, filled with water.

❷ FLAT TRAY, FILLED WITH WATER OR WITH GRAVEL & WATER

To increase humidity by evaporation, you may use a flat tray out of some water-resistant material like plastic, stainless steel or earthenware. This tray should be two to three times larger than the bonsai pot. Make certain that the tray is not wider than the windowsill you

want to put it on!

Put the bonsai on a flat stone or some flat base so that the pot stands over the water surface — otherwise the roots may rot. Fill the tray with water regularly. As the water evaporates, the atmospheric humidity around the bonsai will be increased.

❷a WATERING BY WICKS

To make regular watering unnecessary, you may use cotton wicks. A crochet hook is put through the soil and pot's hole from above. Catch the wick with the crochet hook and pass it through the soil, making sure than an end of 10 inches (minimum) rests under the bonsai pot. A large pot may have three to four wicks.

Put the bonsai pot on its base. Stick the wick tails among the gravel. The wicks will transport water constantly to the roots.

Caution: If your bonsai grows in a more solid soil there may not be enough oxygen at the roots. Do not water too much. If the tips of the leaves begin to dry out, this is the first sign that a plant has a soil too wet for its roots. If you are unaccustomed to noting very slight changes in your plants, it is best to first try this method on a not-too-precious tree.

❸ BONSAI TERRARIUM

A method for extremely dry rooms or a windowsill with a radiator underneath.

Such a terrarium is constructed like a dollhouse, but the material is glass or polyacrylic sheets, bonded together with a silicone adhesive. You need six sheets for the back and two sidewalls, top and bottom; the sixth sheet is a very small strip in front at the bottom. Fill the bottom with gravel.

In such an enclosure—just the front is open—atmospheric humidity inside the terrarium is rather high. The open front allows easy watering and manipulation of the bonsai and sufficient ventilation.

A terrarium like this can also be positioned within a room. Put an aquarium lamp at its top, leaving the light on for 12 to 14 hours a day. Its best to use a timer.

❹ CLOSED VITRINE WITH HYGROMETER

One method successfully used to grow orchids or bromeliads in a centrally heated room is to put the plants in a closed vitrine, or glassed display cabinet. It is also good for subtropical bonsai trees.

The closed atmosphere carries a certain risk: Because of a very high humidity without any ventilation, parasitic fungi, especially Botrytis, may occur.

To prevent this problem, here are two suggestions: Control humidity regularly—best regulator is a hygrometer—and when humidity is higher than 80%, open the doors. Treat the trees monthly with a mild fungicide.

Finally, not all subtropical plants need the same high level of humidity. Fuchsia and pomegranate, for example, prefer less than 60%. When humidity is too high, small drops will occur at the leaves' tips. My advice: Decrease humidity with better ventilation until water-drops no longer occur on your plant's leaves.

EDITOR'S NOTE:

Wilma Swain, living and working in Canada, offers these additional thoughts on the subject of humidity for bonsai:

- A small wall gauge to measure the humidity is helpful.

- Many homes have an automatic humidifier attached to their furnaces. A portable electric humidifier can be installed near the plant-growing area. If you have the space, a room could be made exclusively for your plants and kept at a lower temperature, automatically increasing the relative humidity.

- The texts examined on this topic omit the fact that sitting water in a tray will develop, over time, a slime or other growth—giving an unpleasant appearance and developing an odor. I recommend adding a fungicide or herbicide to the water of any tray used for evaporation.

- It's advantageous to bunch plants together as this helps to increase the surrounding humidity through their own transpiration.

Land and water penjing using crown-of-thorns (*Euphorbia wilii*).

LAND AND WATER PENJING
USING INDOOR PLANTS

HAL MAHONEY

We are very familiar with the meaning of the Japanese word, bonsai. We are less familiar, however, with the Chinese word, penjing. *Pen* means tray and *jing* means scene. The word penjing, therefore, refers to a tray scene or a tray

HAROLD E. MAHONEY, *President of the Bonsai Society of Greater New York, President of the Long Island Bonsai Soceity and 3rd Vice-President of Bonsai Clubs International, writes, teaches, lectures and demonstrates bonsai.*

landscape. The landscape may possess trees, soil, water, rocks, accessory plants, moss and even small figurines.

There are numerous tropical treelike plants that may be used for indoor penjing. A partial list includes the following: *Malpighia coccigera, Serissa foetida, Ficus benjamina, F. neriifolia regularis, Olea europaea, Syzygium paniculatum, Ulmus parvifolia* 'Catlin', *Buxus microphylla* 'Compacta', *Myrtus communis, Cotoneaster* (varieties), *Cryptomeria,* Azaleas (varieties),

Portulacaria afra.

In this article I'll be describing how to construct a land and water penjing.

There are basically three kinds of land and water penjing. One type involves the placement of land on one side of a tray and water on the other.

In such a case, the division of land to water should be uneven to create interest. (Fig. 1)

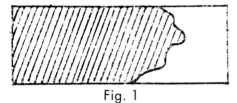

Fig. 1

A second kind of land and water penjing involves land on either side of a stream. In this case, the stream should be located well to one side and not in the middle of the tray. (Fig. 2)

Fig. 2

The third kind of land and water penjing is an island of land surrounded by water. The island should possess an uneven coastline and should be located well to one side of the tray, not in the center. (Fig. 3)

Fig. 3

Fig. 4: mature crown-of-thorns.

Fig. 5: cuttings taken from above.

Fig. 6: planted cuttings.

The unevenness built into all three types of penjing helps to add to the interest of the finished product.

Penjing are usually constructed with outdoor plants but can also be created with indoor plants. An excellent and unusual plant that can effectively be used for penjing is *Euphorbia milii*, dwarf crown-of-thorns. The plants are very easy to propagate, their leaves are small, they are easy to care for, they are virtually disease free, they possess interesting thorns, and their beautiful red flowers are present the entire year.

In order to create a large and interesting penjing, it is necessary to obtain many plants of varying sizes. This is easily accomplished by obtaining two or three older plants from which large numbers of cuttings can be made. A good rooting medium for the crown of thorn cuttings consists of 75% coarse builder's sand and 25% sphagnum peat moss. The cuttings should be of varying sizes. Some should be very small—one to two inches; others—up to six or eight inches. A sufficient number of cuttings can be obtained from just one or two mature plants. (Figs. 4, 5, 6)

The cuttings root easily, so it is not necessary to use rooting hormone. If the cuttings are raised in bright sunlight or under bright artificial lights, they will branch rapidly and will begin to produce their beautiful flowers almost immediately. The cuttings should be ready to use in just one or two growing seasons. With a little luck, you may find all the plants you need at a local nursery.

In planning penjing, proceed basically as you would with an ordinary forest planting. The plan described here would be for a land and water penjing possessing a stream with land on either side.

First, construct a stream bed to one side of the tray, away from the edge. Rocks may be used to outline the stream bed, use the same type to provide necessary harmony but of unequal sizes to create variation. Curve the stream bed so that its source cannot be seen. (Fig. 7)

Place the rocks on wax paper and

Fig. 7

cement in place with a waterproof, quick drying, hydraulic cement such as Thoro's Waterplug. This procedure will allow arrangement of the trees to proceed without displacing the rocks. The wax paper prevents the rock from sticking to the tray and can be removed after the cement dries. Create a curved irregular stream bed to increase interest.

The stream bed should be placed approximately one-third from one side of the tray. The crown-of-thorns "trees" should then be placed so as to create a large, uneven triangle to the left of the stream bed and a smaller triangle to the right of the stream bed. Place the largest tree one-third from one side and slightly to the rear of the tray. The second and third largest trees are then placed to form a scalene triangle when viewed from above. (Fig. 8)

With the three major trees in place, proceed to construct the large triangle around trees 1 and 2, and the smaller triangle around tree 3. (Fig. 9)

Use trees of the same variety to enhance the harmony in the planting. However, use trees of different heights and different trunk diameters. Plant at varying heights above the tray with the tallest trees on the highest hummocks and spaced unevenly. Plant the largest

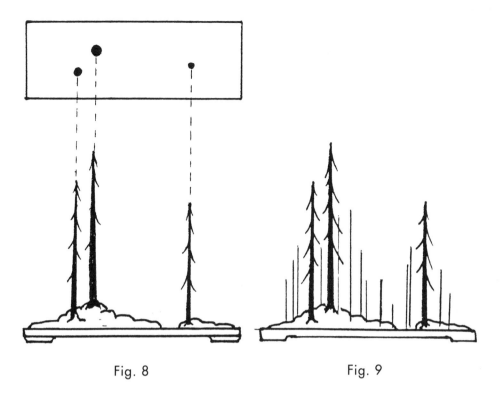

Fig. 8   Fig. 9

trees to the front and smaller ones to the rear to create perspective. The surface topography of the soil should be very uneven and covered with moss to hold the soil in place and give the penjing a well established look (Fig. 9). If the moss is collected from an area of light shade, it will stay green all winter long under strong fluorescent lights. Additional rocks may be placed at varying intervals singly, and in small groups to create further interest.

While any oval or rectangular tray may be used, a white Chinese tray, 23 inches in length, was used in this construction. The tray is unique in itself and serves as a beautiful frame for the green plants, red flowers, moss and rocks.

59

Texas ebony.

TEXAS EBONY

EDITH SORGE

I have grown many bonsai suitable for growing indoors but my loyalty and love are steadfast for the Texas ebony.

The Texas ebony (*Pithecellobium flexicaule*) is a native of southern Texas and Mexico. It is a large broad-leaved shrub, or small tree, usually not attaining a height above 30 feet. It is a rounded top tree with dense, dark green foliage.

EDITH SORGE, *originator of The Bonsai Farm in Adkins, Texas and founder of the San Antonio Bonsai Society, has appeared on television, conducted workshops and classes and published articles on bonsai.*

Texas ebony blooms once a year, generally after a summertime rain, at which time it also drops its ripe seeds. Blooms are golden puffs in spikes followed by beanlike seedpods four to six inches long by one-half to one inch wide, containing many dark reddish brown seeds. Seeds have an extremely hard coating. Leaflets are one-eighth to one-quarter inch, round, arranged feather-fashion on small twigs, usually three to five pairs of leaflets to each two- to three-inch twig. Leaves fold up and droop at night and on dark days, even indoors when under lights.

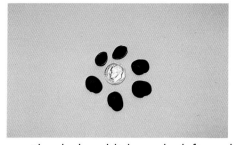 

The dark reddish seeds, left, and the leaves, right, of Texas ebony.

A tropical tree, the ebony will not tolerate temperature below 32° F.

It is one of the easiest plants to work with as it endures neglect and adapts to low light—although full sun is preferred. If underwatered, it will drop all its leaves but within a few weeks, when watered, will put on a new set of leaves.

It is sometimes advantageous to defoliate an older ebony once a year—during the summer—to give it a few weeks resting period. If this is not done it will defoliate itself every few years. First the leaves fold up, then droop and finally turn tan and fall off. It's disconcerting to watch so I find it less disturbing to defoliate it myself. It is even good for the plant to be defoliated prior to potting as it seems to go through this shocking experience much more easily.

When the plant is large enough to make into a bonsai, cut the branches back, decide the height and reduce the trunk height, if necessary. Before wiring, cut the ends off the thorns to keep from injuring yourself. The wire does not have to be wrapped, especially if you use aluminum wire. Wire in the spring and summer and pot during the warm growing season. Style as an informal upright or a cascade if you find one with a low branch. Prune any time during the growing season to achieve the desired shape.

The soil mixture is not critical. I have seen them growing and thriving in black claylike soil. Our soil mixture consists of fine bark, peat moss, coarse blasting sand and Turface. (The only mix which it seemed to resent was bark and perlite, with the perlite being predominant. In this mix the leaves turned a pale green and very little growth was noted.) The trees grow naturally in black loamlike soil. I must deduce that the Texas ebony likes a heavier mixture than many other bonsai.

I usually choose glazed pots for this bonsai since it has such pretty dark green leaves and is not deciduous. I like the blues, greens and whites best.

Texas ebony may be grown indoors very successfully, either near a window or under artificial light. Although it loves full sun, it has been observed in one office on a desk at least fifteen feet from the nearest window. It was not thriving but it was growing. Perhaps lack of fertilizer may have been the reason for little vigorous growth. Fertilize with a well-balanced water-soluble fertilizer once a week or use a time release fertilizer. We use both with ours, wanting them to grow rapidly.

If you have a small plant, place it in a large nursery can, supply water with continuous drip and fertilize two to three times a week. If this method is used it will grow into a large plant in just a few years. If placed in a bonsai pot while small, it will tend to stay small as the pot restricts the roots. Repot every two or three years and root prune lightly.

The only pests we have noted have been scale, which we hand pick from the plants. It is not normally subject to fungus root rot though we treat all of our plants with Benomyl, or similar product, to prevent root rot.

A TROPICAL BLOOMING AND FRUITING BONSAI

MURRAYA PANICULATA, ORANGE JESSAMINE

HEIDRUN HUNGER

Murraya paniculata is a member of the Rutaceae. A native of South China, India, Indonesia and Australia, it is a large evergreen shrub with a sometimes tree-like growth habit. It can reach 10 to 12 feet high. Its leaves have three, and up to nine, broad to ovate round, toothed or toothless, small (one- to two-inch long) leaflets. They are glossy dark green which gives a graceful contrast to the light-colored bark. The bark is used for cosmetic purposes. The white flowers are bell-shaped, fragrant, appearing in clusters. The red fruits contain two seeds.

HEIDRUN HUNGER *is a horticulturist in Leipzig, specializing for the past ten years in tropical bonsai.*

Vegetative propagation is possible through cuttings, but high bottom temperature of 28-30° C (80-86° F) and high humidity are required. Seeds have highest germination rate right after they have reached maturity. The fruits are washed and seeds are separated from the fleshy fruit capsule. Seeds are planted in a mixture of equal parts sand and peat moss (1:1) and placed in a warm location of 20° C (68° F). The soil mixture should be kept moderately moist. Germination takes place in one to two weeks. When the plants reach the two to three leaf stage, the young plantlets are transplanted into two-inch pots with a porous nutrient-rich soil mixture. Filtered sunlight through east or west windows is beneficial. Direct sunlight and southern exposure should be avoided. During the first two years, maintain normal growing conditions: Transplant into large pots and, if necessary, fertilize April to September every two weeks and keep the temperature constant during the entire year at 20° C (68° F). Under these optimal growing conditions the plant will produce its first flowers. It is advisable to remove all faded flowers, since fruit development will weaken the plant.

During spring of the third year of normal growth, the main lead can be wired and bent to a cascading side branch. At the highest point of the cascade a new shoot will become the main lead. Through this technique the lower twigs can be arranged to a graceful *Moyogi*.

Hardwood development starts in the early stages of young shoots. The wood is very brittle and will break easily when shaping the plant with wire. It is better to achieve the desired form through pruning. Another way is to fasten wire under the rim of the pot and slowly bend branches down into the desired position;

the wire will not imbed in the bark of the wood. The pruning can be done year round.

To shape a young *Murraya* bonsai, remove five leaves after the young shoots have developed eight to ten leaves. During that growing period the hardening of the young shoots occurs. With this practice a better-balanced bonsai can be achieved. New shoots will be developed from axillary buds. For a delicate form of the individual branches, prune back to the second leaf of a new shoot after it has developed four to six leaves. Usually at this growing stage flower buds will set. Keep this in mind during the slow process of shaping the bonsai. Over the years the distance for flower buds gets shorter. It is not rare to have 10 to 12 single flowers on one branch. The single flowers last no longer than one to two days. But since not all flowers are open at the same time, the delicate jasminelike fragrance can be enjoyed for a longer term. Shortly after flowering, the small green fruits are visible. Within a half year fruits ripen and turn from green to orange to red, and can reach the size of a cranberry. Not all fruits will reach maturity; most of them will drop off. The few which reach full maturity decorate the shrub with their bright color for about a month. Quite often there are flowers, young and mature fruits at the same time on the plant. This adds a special charm to the bonsai!

Beginning at the fourth growing season, the plant can be transplanted to a shallow bonsai dish. At this time some of the well developed main roots can be pruned back. The plant generally does not have a strong root system, therefore a drastic pruning of the roots is not required. Most of the old soil should be removed. The plant can be easily positioned into the new dish and fresh humus-rich soil should

Murraya paniculata, orange jessamine.

be added. Maintain enough moisture, but do not overwater. The root system should be evenly moist.

Murraya does not respond well to drastic foliage reduction, especially young plants. It can lead to total leaf loss. Full recovery is difficult for the plant. Be prudent with leaf pruning. Older plants have longer internodes between single leaves, therefore leaf pruning is not necessary. 🌳

SUCCULENT BONSAI

C. GLASS & R. FOSTER

To the average plantsman the combination of "succulent" and "bonsai" seems an incredible conflict of terms. Succulents are plants one can grow with little care and bonsai is an art which often demands daily attention. Succulents often (perhaps too often) have a wild, rank, unkempt look, and bonsai is the epitome of meticulous, painstaking, traditional cultivation. But if one looks at the two situations from a slightly different angle, that of overall effect, there are striking similarities.

Some of the best bonsai plants are those that have been naturally dwarfed

Ficus neriifolia grove.

(miniaturized) by the site and conditions under which they grow in the wild. Some succulents in the wild endure virtually the same conditions, and in some cases with a very similar effect. The branches of a succulent tree in the desert are often pruned by hungry, grazing animals or pinched back by frost, and shaped by sun and wind. A succulent's roots are often impeded in their growth by growing in a too-small pocket or rock crevice, or by a lack of moisture in the soil.

The beauty of the succulent bonsai is not only in the graceful or grotesque, contorted shapes, but in the fact that it does not need daily watering! One can leave a succulent bonsai unattended for days or even weeks without major ill effects.

If one is fortunate to obtain a plant which is naturally "bonsaied" by conditions in the wild, its a fairly easy and enormously rewarding job to maintain this bonsai quality in cultivation.

Perhaps the species most ideally suited for this "succulent bonsai" treatment are the various species of "Elephant tree," of the genera *Pachycormus* and *Bursera* or *Commiphora*, but several other plants may be used equally effectively—succulent species of the *Ipomoea*, or Morning Glory tree from Mexico; some of the sedums such as *Sedum frutescens* or *oxypetalum*; some of the mesembs, such as *Trichodiadema densum* or *bulbosum*; and even some cacti, such as old-gnarled specimens of *Opuntia ramosissima.*

Select a shallow container that will not keep the soil wet too long. The best container is the traditional Japanese bonsai pot, not only for its aesthetic appeal and elegant lines, but also because it provides excellent drainage, the proper proportions and because it is made of good clay, high fired and made to last.

Watering requirements are not too different from the average succulent—they can be soaked thoroughly and then, unlike regular bonsai, allowed to dry out considerably before the next watering. The most important element in the care of succulent bonsai is constant and careful pruning and shaping.

Most of the new growth should regularly be removed at least back to the first or second node (or bud). The crown should be opened up by removing much of the inside growth. Suckers on the trunk and underside of the branches should be cut off, and even major branches may be removed to create a more pleasing, artistic effect, but should never be crudely hacked off leaving an ugly stump; remove such branches close to the trunk and the cut will heal in a very natural fashion, enhancing the appearance of the plant. Neither should one prune the plant from just one angle, but repeatedly rotate the plant while working on it, so that the final effect will not be one-sided. Just remember at all times that you are trying to create the effect of a normal tree but in miniature. The result will not, in most cases, measure up to the strict traditions of the true art of bonsai, but it takes much less work, much less training, yet affords very much the same pleasure and satisfaction. 🌳

REPRINTED WITH PERMISSION FROM *THE BONSAI BULLETIN*, THE BONSAI SOCIETY OF GREATER NEW YORK, WINTER 1975, VOL. 13, NO. 7.

WIRING TECHNIQUES

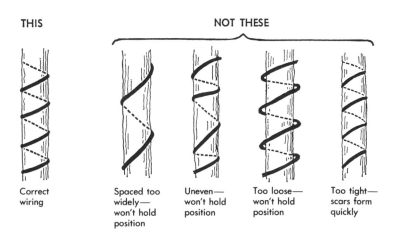

THIS   NOT THESE

Correct | Spaced too | Uneven— | Too loose— | Too tight—
wiring | widely— | won't hold | won't hold | scars form
        | won't hold | position | position | quickly
        | position

Use no. 10 to no. 26 copper wire, depending on branch thickness and stiffness. Soften larger-size wires by bringing to a red heat in a flame and letting cool gradually. Once coiled around a branch, softened wire soon hardens.

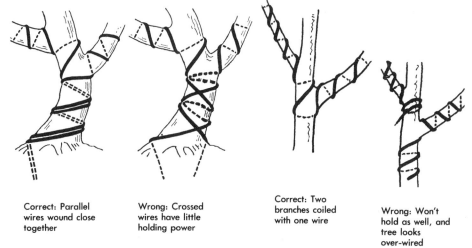

Correct: Parallel wires wound close together

Wrong: Crossed wires have little holding power

Correct: Two branches coiled with one wire

Wrong: Won't hold as well, and tree looks over-wired

Begin wiring at bottom of tree and work upward. If trunk is to be wired, anchor wire ends by pushing down through root ball to a bottom corner of container. Avoid sheathing a tree in wire; good wiring practices give best results with least wiring.

PORTULACARIA

& OTHER SUCCULENTS FOR BONSAI

JAMES J. SMITH

Many people have been growing succulents as bonsai, but not much as been written on the subject. Since succulents are some of the easiest plants to grow and are very tolerant of extremely dry conditions, they are ideal plants for the beginner or for anyone who can't or doesn't water his plants every day. I am not recommending that you neglect succulents. They do respond to good care and if you give them ideal growing conditions, they will reward you by being healthy, good-looking plants. Since they are "survivors" they will live under adverse conditions and tolerate some neglect.

Succulents are plants adapted to survive with less-than-average water supply by storing water in specially enlarged spongy tissue in their leaves, branches, trunk and roots. Some plants use just the roots for storage—such as *Trichodiadema bulbosum*. *Ficus salicifolia* uses its roots, trunk and branches for storage, but *Bursera* uses just the trunk and branches. Jade plant uses the trunk, branches and

JAMES J. SMITH *owns and operates a nursery in Vero Beach, Florida and is an expert on bonsai growing. He writes articles, gives lectures, demonstrations and instruction in bonsai.*

leaves, and *Portulacaria* uses all parts — roots, trunk, branches and leaves. Naturally there are different degrees of succulence. *Ficus* may live for days or weeks without any water, whereas the jade plant may live for months.

The length of time a plant can live without available water depends much on the health of the plant, environment, light, humidity, temperature and wind. A plant grown in low light indoors or full shade outdoors needs less water than one growing in light shadow or full sun outdoors. Indoors, light intensity varies, depending on location, size of windows and distance the plant is from the window. Humidity has a lot to do with how often a plant needs to be watered. Plants give up most of their moisture through their leaves. In high humidity, plants will transpire less than in low humidity.

Although succulents differ from other plants in their ability to withstand drought, requirements are basically the same as for other plants and they must be provided with an environment suitable for their needs. Since most succulents are tender plants they require temperatures above freezing. Most will tolerate extemely high temperatures. Light requirements will vary. The jade plant

Jade plants make excellent bonsai.

will adapt to very low-light levels whereas *Portulacaria* thrives in bright light.

Soil and Potting

Soil for succulents is basically the same as for other bonsai: It must drain well and be heavy enough to support the plant. But since some succulents hold a large amount of water in their leaves and branches, they tend to be top-heavy. Thus the soil should contain a large amount of coarse sand or other heavy aggregate to support the plant. It is usually necessary to wire the plant to the pot until it becomes established. Before potting, let the soil dry completely, then use the dry soil for repotting, especially if any large roots have been cut. Make certain the plant is wired securely in the pot so it can establish itself without any damage to the roots later. Do not water the plant for several days or until the roots that were cut have calloused. If the roots of some succulents are not allowed to callous before watering, they may rot.

Portulacaria needs repotting every year while it is in training; after it has developed to your satisfaction it can be repotted less often. *Bursera*, on the other hand, grows extremely slowly and will not need repotting for ten years or more, depending on the size of the container. Although most succulents can live for many years without repotting, it is a good practice to check the roots each year to determine if the plant is potbound and in need of repotting.

Insects and Diseases

Insects and diseases have not been a problem with any of the succulents mentioned. If you should have to spray, always follow the directions on the label. Most plants tolerate insecticides. *Portulacaria* does not like any pesticide with a petroleum base. If you need to spray any plants that usually will not tolerate a pesticide, test spray a junk plant in a shaded area or, better yet, spray after sundown or on a cloudy day when the temperature is not too high. You may need to wash the chemical off the foliage after a short time. If the plant won't accept the treatment, you may try spraying with soap and water. It is still advisable to spray in the shade and then rinse thoroughly with fresh water from your hose. Use as much pressure as is safe for the plant.

If the plants are small you may wish to dip them a bucket of soap solution and then rinse. If you accidentally spray a plant with a harmful chemical, wash it off immediately with fresh water. Usually no harm will be done.

Propagation

Propagation of most succulents is extemely easy. Many will start from leaf cuttings. A leaf from a jade plant that has fallen to the ground will take root and produce many plants without any effort.

Almost any size cutting can be used; always take cuttings from healthy plants. Let the cut end dry thoroughly in the shade (usually 24 hours is sufficient) before inserting them in moist rooting medium. Another method is to insert the cutting into a dry medium immediately after they are taken from the stock plant.

It is important not to water the cutting until the cut end has dried. the cutting should be placed in a shaded area, protected from strong winds until rooted, gradually moved to brighter light and then fed liquid fertilizer. After they have developed a heavy root system, transplant into training pots. Arrange the roots carefully, keeping in mind the style of bonsai you desire. The rooting medium may be the same soil you use for potting.

Style Training

Succulents can be trained in many bonsai styles. The *Bursera* and jade plant are naturals for informal upright; *Portu-*

lacaria can be trained as a formal upright, cascade or anything in between.

Pruning

Pruning most succulents is slightly different from woody plants. The soil should be allowed to become very dry. When removing a branch, do not make the usual concave cut. Instead, cut flush with the trunk. Usually after the cut has dried, the thin layer of tissue protruding from the trunk will fall off. If not, another cut closer to the trunk may be necessary. Since succulents are not woody until they are very old, pruning with bonsai scissors is very easy. Pinching can be done with the fingers and must be done faithfully in order to develop a fine bonsai.

Portulacaria grows very fast. It is necessary to pinch every week during the growing season. When pinching, always keep in mind that new growth starts at the base of the leaf. If you want vertical growth (top of tree), pinch to vertical leaves. You can always wire the branch and twist it in the direction you want it to grow. Leaf pruning can be very effective on the jade plant; regular weekly pinching on *Portulacaria* will reduce them to one-fourth of their regular size. Since their leaves are normally smaller, *Portulacaria* is ideal for mame, or miniature, bonsai.

Wiring

Wiring is the same as with other plants, except that a thinner wire can be used on succulents. If you should crack a branch while bending, return the branch to its original position without removing the wire. It will heal in a short time. The branch can be very carefully bent later. When removing wire from succulents, it is safer to cut the wire in short pieces than to remove it in one piece, as is often done with other woody plants. I use copper-covered aluminum wire. It takes less effort to wire and I'm less likely to damage the plant.

Training Portulacaria in Bonsai Pots

Since I grow bonsai commercially, it became evident that to produce a respectable looking plant at a reasonable cost, I would have to reduce the growing time to cut the labor costs.

Portulacaria provides possibilities for commercial bonsai because it may be started from leaf cuttings and is fast-growing. Instead of using rooted plants, start with unrooted cuttings and root them directly in the bonsai pots. Some styles, such as group plantings, can be developed in one year or less, while a formal upright will take longer.

The potting soil I use here in Florida is a general nursery soil made up of Florida peat and cypress chips that has been screened through a half inch mesh to remove the larger particles. (Use these larger particles as bottom soil in deeper pots.) Add 50% coarse sand and a slow-release fertilizer with minor and trace elements. Using dry soil allows me to plant the cuttings immediately.

Before planting the cuttings in the pot, do all necessary pruning and wiring because it will be impossible to do any wiring later on—until they are rooted. Our objective: Shorten the training time. If you have a large supply of stock material, much of the training can be done while on the stock plant. Choose your cuttings carefully so by the time they root you will be well on your way to having a bonsai.

After your cuttings are prepared, insert them in the bonsai pots that have been filled with soil. The potted cuttings are placed on a bench under 50% shade cloth.

One or two days later they are watered for the first time. The soil is then kept slightly moist until they have rooted. After they have rooted well, feed with liquid fertilizer regularly for fast development.

*REPRINTED WITH PERMISSION FROM **FLORIDA BONSAI MAGAZINE**, 1989, NO. 4.*

Ficus benjamina 'Exotica' showing aerial roots.

FABULOUS FICUS

James J. Smith

The genus *Ficus* has many species suitable for indoor bonsai: *F. benjamina, F. benjamina* 'Exotica', *celebensis, citrifolia* (short-leaved fig), *microcarpa* (green island), *philippinensis, salicifolia* (willow-leaf fig), *triangularis, retusa* (Chinese banyan), *rubiginosa* (rusty fig), *aurea* (strangler fig).

As well-established trees, the above species are able to withstand a severe drought. Aerial roots are easily developed on all of these except *F. celebensis* and *F. triangularis*. Most will tolerate a low light (indoor) environment and can be trained with the clip and grow method. Repotting is easy because drastic root pruning is safe for most *Ficus* spp.

Location

Light is usually the limiting factor for growing plants indoors. Although *Ficus* will tolerate low light, a location near a window with bright light will produce healthier and more compact foliage. A south window is the best choice. The second—providing that none of the widows are shaded by trees or other obstructions—is an east or west exposure. If this is not possible, grow lights may be substituted. (See articles on *Growing under Artifical Lights*, page 4 for detailed information.)

Watering

Most potted plants prefer a moist soil and *Ficus* is no different. A good rule for watering any potted plant is to water thoroughly (soak the soil) until all the air is removed from the soil. Then let drain until water ceases to drip from the drainage holes. Do not water again until the soil approaches dryness but is still moist.

Schedule your watering to meet the needs of the specific plant—which may be every day, every other day or once a week. If a plant is moved, the watering schedule may have to be changed. Plants growing in bright light will need more watering; a tree that has been defoliated will need less—until new growth appears. More water is needed when the plant is growing; you'll need to schedule watering according to the season. And keep in mind that individual plants grown in different soil mixes will have different watering requirements.

Insects and Disease

Ficus are relatively free from insects and disease but no plant is completely immune. Learn to identify plant problems. Inspect your bonsai each time you water them and if you suspect a problem, examine the leaves in bright light—especially the undersides using a 5X magnifying glass. Some pests are as small as 1/100th of an inch long. Most problems show up on the leaves: leaf curl, discoloration, leaf spot, powdery growth, wilted leaves and leaves chewed by insects. The woody parts of the tree should also be inspected for insects and for rot.

It is important to identify the pest and then use the correct treatment as soon as possible. When using any pesticide, always follow the directions on the label at the strength directed. If in doubt, ask a knowledgeable grower or your County Extension Agent.

Fertilizing

Any all-purpose fertilizer that contains micronutrients is recommended. Liquid fertilizers are quick-acting and must be applied more often than dry fertilizers. During the growing season, apply weekly if growing outdoors; monthly if indoors. Insect-free foliage that is paler than normal in color indicates that the tree needs fertilizer, whereas darker green foliage shows it does not.

Granular fertilizers are available in quick-acting and slow-acting formulas and may be applied either to the surface of the soils or mixed in with the soil when potting the tree. When using quick-acting fertilizer, always make sure that it is distributed evenly over the soil surface so that it does not burn the roots. And never use more than the directions indicate. Care in watering is needed so as not to wash the fertilizer from the soil.

Some slow-release fertilizers are available in the form of tablets, sticks or briquets. These are placed below the soil surface and release their nutrients as soil conditions change. Some are governed by the bacterial action in the soil, others are made available to the tree by the amount of water given the soil. These products will probably last anywhere from two months to two years, depending on the product selected and your growing and watering conditions.

Organic fertilizers are slow release and depend on bacterial action to make them available to the plant. They need to be applied about every two months.

All dry fertilizers can be mixed or supplemented with liquid fertilizers. Always use as directed. Again, if in doubt, seek advice. Plants growing in full sun need more fertilizer than those growing in shade; adjust your fertilizing schedule accordingly.

Repotting Ficus

Bonsai need repotting when the pot is filled with roots and, if not done, the excess roots will cause the tree to rise in the pot.

A *Ficus* bonsai growing outdoors in South Florida may need repotting each year, whereas the same tree growing indoors in New York may not require it for five or more years.

Spring and summer are the optimal times to pot *Ficus*, although with proper care and good health, it may be done any time of year. Most healthy *Ficus* will tolerate drastic root pruning.

When repotting, shorten all branches as needed to maintain the desired shape of the tree. Remove all leaves and terminal buds (an exception might be a weak branch or one that needs additional growth). The steps and soil and watering requirements for repotting *Ficus* are the same as for other bonsai.

Training

The size and shape of the *Ficus* tree can be maintained by pinching the terminal bud of any branch when it reaches the desired length. Continue pinching throughout the growing season, leaving one or two new leaves. You can direct new growth by pinching back to buds aimed in the direction you wish new growth to take. Wire can be used to change direction of heavier branches.

Propagation

Ficus are easy to propagate from cuttings and air layering. One method that is of interest to bonsai growers is root cuttings. The procedure is simple.

A root system that has been removed by cutting horizontally across the entire root ball of a large *Ficus* should be planted intact so that the cut ends of the roots are even with the surface of the soil. Keep in a partially shaded location until new growth appears at the cut ends of the roots. Then move it to a more sunny location. Water requirements are the same as for other bonsai.

A forest or group planting can be created from the vertical stems growing from the root cuttings. As the planting develops, remove all the unwanted growth, leaving only the trees you want to keep. Trim the trees so that each one of them is a different height. The diameter of the trunk will determine the height of the tree; thickest trunk will be the tallest, thinnest trunk, the shortest. After the planting has developed sufficiently remove enough roots so that it can be planted in a shallow tray. The root system that was removed can be used to start a new group planting.

Aerial Roots of Ficus

Those of us who have had occasion to be in the tropical areas where figs grow have been impressed with their natural tendency to grow aerial -or pillar roots. The name given to any type of tree that exhibits aerial roots and forms a grove is Banyan.

Most figs naturally tend to be upright trees, covering large areas of ground by spreading laterally. They can do this by dropping aerial roots and transforming low-arching branches into tree trunks that are all interconnected. In many areas one can see *individual trees* that cover *acres* of ground by this process.

Aerial roots are very attractive and will form naturally on bonsai plants that are grown in high humidity and shade; and from older, more mature wood. Articles have been written on how to encourage *Ficus* to develop aerial roots. Many techniques will work but not all in a reliable fashion. The most successful way to develop an aerial root is to graft a seedling. Graft onto an area of the mother plant that would be most attractive.

CHINESE ELM BONSAI

RANDY BENNETT

Growing Chinese elm (*Ulmus parvifolia*) for indoor bonsai is very rewarding. There are several reasons which account for its success: First, its adaptability to a wide range of climatic conditions; second, it develops quickly, allowing the creation of good bonsai in a relatively short time; third, Chinese elm is easy to obtain (readily available from landscape and garden nurseries due to its popularity as a landscape tree); fourth, it is resistant to disease.

Characteristics

Chinese elm is often referred to as the Chinese evergreen elm. This is due to its tendency to retain its leaves through much of the winter in southern regions where temperatures are mild. In fact, some of the Chinese elm cultivars may retain leaves for several years before losing them.

Ulmus parvifolia is a subtropical tree native to parts of China, Korea and Japan. It is recognized by its nearly smooth bark that exfoliates in thin scalelike layers. The leaves are one and a half to two inches long, obovate and serrate. It flowers and fruits in the fall. Trees may reach heights of 60 feet (18 meters) if given enough time and space. Chinese elm has a natural growth pattern which forms multiple branches from the same point on the trunk and continues to ramify into the basic broom style. Although this is its natural tendency, it in no way resists being styled into any bonsai design.

Chinese elm and its cultivars develop thick root structures quickly and give young trees the desirable appearance of age. The thick roots are also very pliable and make this species one of the easiest to work with when creating root-over-rock and root-on-rock designs.

Ulmus parvifolia may be propagated from seed, branch cuttings, root cuttings, or air-layering. I have successfully propagated Chinese elm from branch cuttings over an inch in diameter. Some of the cultivars are more difficult to propagate from cuttings, but if you can locate a source for Hormex or Hormodin rooting powders[1], the job will be a simple one.

Chinese elm started from seed will develop a long single taproot. If the taproot is cut back drastically when the seedling is two to three years old, it will throw out radial roots which will thicken quickly to form a good base. Cuttings and air-layers develop radial roots at the outset and only need routine pruning to encourage their proper development.

RANDY BENNETT *is a school teacher and artist with a keen interest in bonsai. He is editor of the* **New Orleans Bonsai** *magazine.*

Ulmus parvifolia 'Seiju' forest planting using branch cuttings in root-on-rock style. Tallest tree is five and a half inches tall and three years old.

Cultivars

There are numerous cultivars of Chinese elm. Each cultivar has unique and interesting characteristics and all exhibit desirable qualities for use in bonsai. Some of the more easily obtainable cultivars are listed below.

'Catlin'— A dwarf cultivar with smooth bark. Leaves are thick, dark green, obovate, crenate and shiny; from one half to three fourths of an inch long. Good branching characteristics. Moderate to fast growth. Trunk thickens slowly.

'Cortica'— A semidwarf variety whose bark is rough and corky with deep vertical furrows. Fast growing shoots may exhibit corky wings. Leaves are rough, obovate and are from one to one and a quarter inches long.

'Drake'— A semi-dwarf elm with smooth bark which exfoliates when mature. Dark green, obovate, crenate leaves from one to one and a half inch in length. This species exhibits a weeping habit with good branching. Fast growing.

'Frosty'— A smooth-bark, dwarf, shrub-type cultivar with leaves variegated only along the edges. Leaves are obovate, serrate and from one and a half to two inches long. Good branching characteristics. Medium growth.

'Hokkaido'— A miniature cultivar with extremely thick, rough and corky bark. Leaves are orbicular, crenate and one sixteenth to one eighth inch in length. Excellent twigging. Slow growing.

'Seiju'— A dwarf cultivar which develops very corky bark with deep vertical fissures even on young trees. Leaves are obovate, crenate and from one quarter to one half inch long. Excellent twigging. Fast growing.

'Stoney's Dwarf'— A dwarf variety similar to 'Seiju' but with slightly larger leaves from one half to three quarters of an inch in length. Bark is more knotted in appearance and without vertical fissures. Good twigging. Medium growing.

'Suberosa'— A semidwarf cultivar with very rough corky bark. On a tree with a two and a half inch diameter trunk, about an inch is bark. Leaves are rough, obovate, crenate and from one to one and one half inches in length. Good twigging. Medium growth.

'Yatsabusa'— An unusual dwarf variety with rough bark that forms much thinner bark than the other cork bark cultivars. Leaves are spathulate, crenate and from one half to three quarter inches in length. This is a weeping cultivar with excellent, very delicate, almost lacelike twigging. Moderate growth.

Soil

Ulmus parvifolia originated from the lower elevations (under 1,000 feet) of southern China along the river regions. It thrives in a soil high in organic matter. In bonsai, however, soil requirements must balance with good drainage, especially in the indoor environment. *The number one cause of indoor plant death is overwatering in soil that has poor drainage.* Therefore a proper soil mix is the single most important element in indoor bonsai culture.

The more ingredients you use, the more difficult to discern what changes to make if your soil mix has problems. You must, therefore, keep the mix simple. I use two ingredients for my indoor Chinese elms: 1. pine bark mulch sifted through a quarter inch screen. This provides nutrients as it breaks down and retains moisture in the soil. Avoid using peat moss or any other organic material which is too fine. It will become packed in the soil, inhibit air circulation, retain too

much water and thus cause root rot. 2. haydite. It is the "sand" used in the manufacture of cinder blocks. I sift it through an eighth inch screen and eliminate anything which falls through a sixteenth of an inch screen. It has all the qualities of an excellent drainage material.

The drainage material is even more important than the organic. Avoid materials which have a smooth or polished surface. They do not provide surface adhesion for water and air. Also avoid drainage materials which are *flat* such as some of the expanded clays. These will tend to pack down in layers with each watering and stratify the soil, preventing air circulation and poor water adhesion. Use a material which is angular and has a rough surface or one which is filled with tiny holes (as haydite).

I provide the elms I keep indoors with a 60% haydite/40% pine mulch mixture. These proportions work excellently given the amount of humidity, light, temperature, air circulation and the species. The mixture that is right for you will depend on the climate in your area and the environment in your home. But do not spend a great deal of time worrying over it. It is the nature of *Ulmus parvifolia* to adapt to conditions which may be too wet or too dry for other species.

Humidity

The warmer you keep your home during winter, the greater the need for supplemental humidity. The means of increasing humidity for Chinese elm is standard for all bonsai. (Editor's Note: Please see the article on Humidity, page 52.)

Temperature

Ulmus parvifolia is a subtropical tree. Subtropicals typically need a winter temperature of between 41° and 54° to provide the needed *resting period*[2] . Here again is where the Chinese elm shows itself to be the ideal tree for indoors. It will tolerate winter temperatures of between 64° and 72° while still maintaining its period of rest. Just remember that during the winter the Chinese elm needs at least six weeks of rest and you should help it out by keeping the room as cool as possible.

Although Chinese elm is a subtropical species it is hardy to Zone 5. Zone 5 indicates that the average low winter temperature for that region is between -10° and -20°F . Some of the cultivars mentioned, however, are not as winter hardy as *Ulmus parvifolia.*

Watering

The frequency of watering will be determined by the soil mix and how quickly it dries out. The best practice with Chinese elm is to allow the soil to dry out somewhat and then water again. Otherwise its care is the same as for other bonsai. (EDITOR'S NOTE: Please see the article on Watering, page 28.)

Location & Light

Three things must be considered when siting Chinese elm indoors: the cultivar, air circulation and light. Which cultivar you are growing will dictate to some extent where it must be placed. The smaller the leaves, the more light is required by the tree. So if you are growing *Ulmus parvifolia* 'Hokkaido' you will need as much light as possible. Conversely, the larger the leaf, the less light required for proper health. The variegated cultivars also require less light than their counterparts.

Air circulation is essential to the health of Chinese elm indoors. Air movement allows the exchange of carbon dioxide with the cells inside the leaves. Your elm uses the energy from light to split water so that the hydrogen atoms can bond with the carbon dioxide to form a carbohydrate molecules which the elm uses for food. If the air circulation is poor where your elm is located, it

may not be getting the carbon dioxide it needs. *Thus, your elm's health and its ability to produce food is more often affected by a lack of carbon dioxide than light.*

Your elm will use light to convert carbon dioxide and water into sugar. The sugar that is manufactured is the stored energy that feeds the tree. This photosynthesis can only be performed in certain light and is dependent upon the level of carbon dioxide present in the air, amount of moisture and temperature.

Although the manufacturing of food can only take place in light, your elm can use its stored energy anytime—including at night! This can affect your elm in that it will begin using its stored energy if not receiving enough light. When the tree has used all its stored energy, it begins to consume itself. This will be evidenced through pale leaves and weak, spindly growth. A lack of light may also cause yellowing and dropping of leaves, particularly at the bottom of the tree. Providing artifical light to maintain the health of your Chinese elm is a simple task, but there are several things which need to be kept in mind.

First, your elm will only utilize certain wavelengths of light. The red, blue and violet wavelengths are responsible for plant growth. Second, incandescent light bulbs are not suitable for providing light. Their light does not have the necessary spectrum for plant growth and the heat they emit will damage your tree.[3] Third, if you use artificial light to supplement exisiting natural light, exposure should be between six and eight hours. Fourth, where artificial light is the only source, the duration of exposure should be between 10 and 16 hours. Automatic timers insure consistent exposure even when you are away from home.

I have my Chinese elms in my living room in three different locations. I utilize a row of east-facing windows using a combination of natural light and about six hours artificial. A north-facing sliding glass door is used where there is only indirect light and so I rely on fluorescent lights more heavily—about 10 hours. A third site is a brick planter extending out from the wall by the front door. This area receives no natural light so fluorescent light is the only source—for a duration of about 16 hours.

Once you have selected your site, there are a couple of ways to improve the lighting. First, keep the windows clean! Dirty windows can cut down the transmission of light by as much as 40%. Second, increase light reflection by painting white the surfaces that surround your tree—such as shelves, side panels and the area behind any artificial light. You might also try a light-colored curtain behind your trees (which would also serve to block the hot, dry air from heating ducts in winter, and provide a cooler microclimate for your trees).

Do not forget to move your trees around. Rotate them from site to site if they are in areas of diverse light and temperature. That way each tree benefits from its "fun in the sun." Also remember to turn your trees if you are not using any artificial light. Otherwise the trees will tend to grow toward the window and you may lose foliage or even entire branches.

Styling

The natural growth pattern of Chinese elm is in the broom style with no central leader. However, this species can be styled into other bonsai designs. It develops fine twigs and branches and beautiful canopies of tiny delicate leaves.

Wiring is best done in early spring before buds begin to swell. Be sure to keep an eye on wired branches as initial spring growth is quite rapid and branches may be damaged by leaving the wire on too long.

General pruning can be carried out any time of year. If, however, you are removing a major branch, wait until early spring before buds swell.

Fertilizer

Chinese elm responds well to any of the typical fertilizers recommended for bonsai. The best practice is to vary the type of fertilizer. No two fertilizers are exactly alike. Some may contain certain minerals or trace elements absent in others, even though the primary ingredients may be of the same type and proportion. Fertilizer may be applied at any time during the growing season, April to September, but remember that your elm must undergo a resting period so avoid fertilizing during late fall and winter.

Containers

The container used for *Ulmus parvifolia* will depend upon the bonsai style in which it has been trained. They look equally at home in glazed or unglazed containers. I tend to favor oval to rectangular containers because elms develop such soft rounded canopies and delicate, airy foliage. If a glazed container is used, a shade of green complements the color of the foliage. Glazed containers may also be useful if you have a problem with the soil drying out too quickly. The glaze helps to slow evaporation through the pottery surface.

Diseases and Pests

The number-one cause of indoor plant death is overwatering. Number two is underwatering. Number three, too much fertilizer. After that, it is a toss-up for which comes next—not enough light or too much light, temperatures that are too high or too low, too much humidity or not enough.

When any of these things happen, alone or in combination, the tree is weakened. It is in this weakened state that a tree is susceptible to insect attack.

Ulmus parvifolia 'Yatsabusa' trained in the informal upright style for one year and then placed with the roots over a rock for one year. The tree was taken from a branch cutting and stands six inches above the glazed container.

Seek to prevent the cause of insect infestation and not simply the symptons ridding of the infesting agents.

Ulmus parvifolia is highly resistant to disease and pests. But if chemical treatment of your Chinese elm is necessary, you may use any of the usual insecticides for treating elms, but with this caution: Do not use at the recommended strength on the various cultivars. The foliage of most of the cultivars is too sensitive to normal doses of Diazinon and may defoliate the tree and cause severe damage or even death. Malathion or insecticidal soap may be used without worry.

Insects and diseases will usually attack a plant where it is weak or damaged. It is therefore important to keep the tree "clean" by removing dead leaves and dead branches and twigs. Areas such as these create ideal breeding places for pests. 🌱

[1] A source for both is Mellingers at 2310 West South Range Rd, No. Lima, Ohio 44452. Rootone is available at larger plant and garden centers.

[2] A dormant or rest period may be induced by combining a lower temperature with a shorter day length. Bonsai grown under lights benefit from the reduction of light from 16 to 12 hours a day and then gradually increasing the light back up to 16 hours a day. (EDITOR'S NOTE: For details on dormancy and resting periods, please see article *Resting Periods, Light Effects and Indoor Bonsai*, page 36.)

[3] EDITOR'S NOTE: Not completely true. See articles on light. Use bulbs 25 watts or less.

151 PLANTS
FOR
INDOOR BONSAI

SIGMUND DREILINGER

The following list provides cultural requirements, and guidelines for winter care, for 151 bonsai. The list of plants has been compiled from many sources and the cultural requirements derived from the experience of many bonsai growers. These are not hard and fast rules, and certainly may vary under your conditions. Some bonsai are deemed "easy" to grow and are so marked; others are "difficult" and are so indicated.

Light, humidity, fertilizer and watering needs during rest periods have been discussed elsewhere in this handbook — please refer to those articles for details. The following codes are used in the plant list for winter care recommendations.

WINTER CODES:

CODE #1: House temperature 55° night to 75° F day.
Good light needed, not necessarily sunlight.

CODE #2: House temperature 60° night to 75° F day.
As much sun as possible.
Supplement by fluorescent lamps burning for 14 to 16 hours.

CODE #3: House temperature cool 50° to 55° F night to 65° F day.
Needs good to very good light.

CODE #4: House temperature cool to cold; 40° to 50° F night to 60° to 65° F day. Very difficult to grow in average home.

ACACIA BAILEYANA—**GOLDEN MIMOSA**—Code 3
From Australia, has pinnate leaves and fragrant yellow flowers. Needs a cool winter temperature. Repot every two to three years.

A. DEALBATA—Code 3
Keep moist, water when top half inch is dry.

A. FARNESIANA—**SWEET ACACIA**—Code 2
Fragrant flowers. Can take warmer temperatures than *A. baileyana.*

ALBIZIA JULIBRISSIN—**MIMOSA TREE**— Code 3.
Bipinnate leaves, flowers are pink pompons – like powderpuffs. Bright light to induce flowering. Needs a cool winter. Difficult.

ARAUCARIA ARAUCANA (A. IMBRICATA)—**MONKEY PUZZLE TREE**—Code 1
Tolerates some dryness and winter heat. Keep moist, not wet. Slightly acid soil. Peat moss is good. Let become potbound.

A. HETEROPHYLLA (EXCELSA)— **NORFOLK ISLAND PINE**—Code 1
Not a pine. Needlelike leaves, wheel spoke branches. Culture as above.

ARDISIA CRENATA—**SPEAR FLOWER**—Code 2
Keep moist. High light intensity helps produce white flowers which become red berries.

A. CRISPA—**CORAL BERRY**—Code 2
Culture as above.

A. JAPONICA—**MARLBERRY**—Code 2
Culture as above.

BAUHINIA BLAKEANA—**HONG KONG ORCHID TREE**—Code 1
Large two-cleft leaves; can reduce. Purple flowers resemble orchids in late winter and spring. Water sparingly in winter. High light helps flowering.

B. PUNCTATA—**RED BAUHINIA**—Code 1
Dwarf variety. One inch leaves and white flowers.

B. VARIEGATA 'CANDIDA'—**ORCHID TREE**—Code 1
White flowers

BOUGAINVILLEA GLABRA—**PAPER FLOWER**— Code 2
A vine which develops a rough-bark trunk with age. Very showy bracts with small inconspicuous flowers. Many color varieties. Water well in summer. Keep on dry side in winter to induce flowering. Too much water or cold will cause leaf drop.

BRASSAIA ACTINOPHYLLA (SCHEFFLERA ACTINOPHYLLA)—**AUSTRALIAN UMBRELLA TREE**—Code 2
A vigorous grower that takes reduced light. Tolerates heat. Has large leaves that will reduce. Keep moist. Train by grow and clip. Easy.

BUCIDA BUCERAS—**GEOMETRY TREE, BLACK OLIVE**—Code 1.
Larger leaves than *B. spinosa*; thorny, leathery leaves. Keep warm. Transplant only in hottest months – July & August. Give high light and keep moist.

B. SPINOSA—**BLACK OLIVE**—Code 1
A smaller leaved species with zigzag branches. Keep moist. Cut back and raise a branch to be terminal to get interesting trunk. Needs high humidity. Root prune only July or August. More desirable for a bonsai than *B. buceras.*

BUDDLEIA INDICA (NICODEMIA DIVERSIFOLIA)—**INDOOR OAK**— Code 1
Wavy lobed leaves reduce easily. Pinch

Serissa foetida 'Variegata'

PHOTO BY CHRISTINE M. DOUGLAS

frequently to induce branching. Keep warm, moist. Bright light.

***BULNESIA ARBOREA*—VERA, VERAWOOD**—Code 2
From Columbia and Venezuela, S. America. Pinnate leaves, brilliant yellow flowers. Reduce watering in winter.

***BURSERA MICROPHYLLA*—Code 1**
Culture as above. Slow grower.

***B. SIMARUBA*—GUMBO LIMBO**—Code 1
Peeling light brown bark; compound dark green leaves. Large diameter cuttings root very easily. Tolerates high temperature and dryness. Fast grower.

***BUXUS HARLANDII*—CHINESE BOX**—Code 1
Tolerates heat and dryness.

***B. MICROPHYLLA JAPONICA*—JAPANESE BOX**—Code 1
Evergreen leaves; acid soil.

***B. M.* 'COMPACTA'—KINGSVILLE BOX**—Code 3
Needs winter rest; don't overwater. Fertilize sparingly, keep on cool side. Very slow grower.

***B.M. KOREANA*—KOREAN BOX**—Code 3
Culture as above.

Juniperus chinensis var. *sargentii.*

CALLIANDRA EMARGINATA—**DWARF RED POW-DERPUFF**—Code 1
 Pinnate leaves; flower buds like pink raspberries. Acid fertilizer and soil. Trunk thickens slowly. Keep moist. Needs high light to flower.
(* There is some confusion in the literature about the names of the dwarf powderpuff. Some sources say it is *C. haematocephala* 'Nana'.)

C. HAEMATOCEPHALA—Code 1
Larger than above. Culture as above.

CAESALPINIA PULCHERRIMA—**FLOWER FENCE**—Code 2

Pinnate, feathery foliage. Bright yellow flowers in panicles. After seed ripens, the tree reblooms. Don't overwater. Need high light intensity. No wet feet!

C. MEXICANA—**CURLYPOD**—Code 2
Culture as above.

CAMELLIA JAPONICA—**CAMELLIA**—Code 3
Must have cool nights to set flower buds. High humidity, acid soil. If kept too warm the buds will drop. Wood is stiff and brittle. Wire carefully. Difficult.

C. SASANQUA—Code 3
Culture as above. Difficult.

CARISSA GRANDIFLORA—NATAL PLUM—Code 3

Thorny bush from S. Africa. High light needed to flower. Has red, edible fruit. Keep moist. Will tolerate higher temperatures. Resents root pruning.

C.G. 'NANA COMPACTA'—DWARF NATAL PLUM—Code 3

A better variety for bonsai. Culture as above.

CASSIA AUGUSTIFOLIA—Code 1

Has narrow leaves. Prefers cool, but can tolerate higher temperatures.

C. MARILANDICA—SENNA—Code 1

Has compound leaves with yellow pealike flowers. With diffused light it flowers freely. Culture as above.

C. RENIGERA—BURMESE SENNA—Code 1

Has pink flowers. Culture as above.

CHAENOMELES JAPONICA—JAPANESE FLOWER-ING QUINCE—Code 3

Many-stemmed shrub. Needs cool temperature and bright light to induce flowering. Cold rest period in winter, water moderately. High temperatures can be offset by high humidity or misting. Difficult.

CHAMAECYPARIS PISIFERA 'PLUMOSA'—SAWARA FALSE CYPRESS—Code 4

Needs cool temperatures to 65–70 degrees F with cold winter temperatures of 45 degrees F. Give high light and keep branches open by careful pruning for maximum air flow. Keep on dry side. Very difficult.

C. P. 'NANA'—Code 4

Conditions and culture as above.

C.P. 'AUREA'—Code 4

Conditions and culture as above.

C. P. 'SQUARROSA'—Code 4

Conditions and culture as above.

CHRYSANTHEMUM FRUTESCENS—MARGUERITE —Code 3

Needs bright light. Provide cool temperatures in winter (45–50 degrees F) to bloom.

CINNAMOMUM CAMPHORA—CAMPHOR TREE—Code 2

Leaves are evergreen, aromatic, smell of camphor when crushed.Can be reduced. Trunk has fissured look. Grows easily but keep pH neutral to slightly acid.

CITRUS species—many varieties—all Code 3

Keep slightly dry, give bright light. Need a cool winter temperature and acid fertilizer. Supply trace elements occasionally. Water more when in active growth. White fragrant flowers. Varieties: Seville orange, Meyer lemon, Grapefruit, Ponderosa, Otaheite orange.

x CITROFORTUNELLA MITIS—CALAMONDIN—Code 3

CLERODENDRUM THOMSONIAE—BLEEDING HEART, GLORY BOWER—Code 1

A vine with red flowers from W. Africa. Pinch new growth to maintain shape. Long flowering period. Water well.

COFFEA ARABICA—COFFEE—Code 1

White flowers followed by red berries. Needs high light and humidity to induce flowers and fruit. Do not allow to dry.

COCCOLOBA UVIFERA—SEA GRAPE—Code 1

Stands heat. Water when top of soil is dry, but do not overwater. Fruit is edible and makes a sweet jelly. Large leaves can be reduced.

CONOCARPUS ERECTUS—**BUTTONWOOD**—Code 1
Roots very easily in plain water in bright sun. Temperamental indoors. Will drop leaves in a draft or cold temperature. Give high light, high humidity and high temperatures and it will thrive. Breaks new growth easily on old wood. Collecting is prohibited on public lands. Moderately difficult.

CONOCARPUS ERECTUS SERICEUS—**SILVER BUTTONWOOD**—Code 1
Has silvery gray leaves. Culture as above.

COTONEASTER MICROPHYLLUS THYMIFOLIUS—Code 3

COTONEASTER MICROPHYLLUS '**COCHLEATUS**'—Code 3
There are many varieties and cultivars of cotoneaster. All have small leaves, many have white flowers and red berries. Give cool conditions and a winter rest. Good for mame bonsai. Difficult.

CRASSULA ARGENTEA—**JADE PLANT**—Code 2–3

C. ARBORESCENS

C. SCHMIDTII

C. COOPERI

C. arborescens can take higher temperatures. Allow a dry cool rest in winter. Fertilize with a low nitrogen fertilizer. All need a very well drained soil. Water when soil is almost dry.

CRYPTOMERIA JAPONICA '**COMPACTA**'—**JAPANESE CEDAR**—Code 3
Supply high humidity and cool temperatures. Train so that air and light penetrate between the branches. Difficult.

C.J. '**TANSU**'—dwarf—Code 3
Culture as above.

CUPHEA HYSSOPIFOLIA—**FALSE HEATHER, JAPANESE MYRTLE**—Code 1
Small, very narrow leaves. Pink to purple flowers in profusion. good light. Keep moist, do not allow to dry. Easy. Good for small bonsai.

CUPRESSUS ARIZONICA—**ARIZONA CYPRESS**—Code 3
Difficult indoors unless you can supply high light and cool to cold temperatures. Need 40°F in winter. Very difficult.

C. MACROCARPA—**MONTEREY CYPRESS**—Code 3
Conditions and culture as above.

CYTISUS RACEMOSUS—**BROOM**—Code 3
A shrub or hedge plant. Needs a cold rest in winter to flower. Give a bright light and alkaline soil and it will produce many yellow flowers.

EHRETIA MICROPHYLLA (CARMONA MICROPHYLLA)—**FUKIEN TEA**— Code 2
Can take 75 degrees F temperature. Shiny green leaves. Needs bright light to produce white flowers, followed by green berries that turn red when ripe. Keep slightly moist but supply good drainage. Does not like wet feet.

EUGENIA UNIFLORA—**PITANGA, SURINAM CHERRY**—Code 1
Shiny, evergreen leaves, fragrant white flowers, edible red fruit – like miniature pumpkins. Needs very good light, slightly acid soil. Keep moist. Use acid fertilizer. Easy.

E. BRASILIENSIS—**BRAZIL CHERRY**—Code 1
Conditions and culture as above although it can take higher winter temperatures.

EURYA JAPONICA—**JAPANESE ELDERBERRY**—Code 2
Needs warmth and moisture and good drainage. Dark green leaves, black fruit.

Left to right: Rosemary, citrus and *Serissa* summering outdoors.

FICUS

All varieties are good subjects for indoor bonsai. They can survive lower light levels. Some produce a dense canopy of leaves and aerial roots. Among the many varieties the following are recommended:

F. AUREA—**STRANGLER FIG**—Code 1
Develops aerial roots. Leaves reduce easily. Can take heat and some dryness in the house.

F. BENJAMINA—**WEEPING FIG**—Code 1
Can develop aerial roots. Leaves reduce. A somewhat weeping habit.

F. DELTOIDEA (F. DIVERSIFOLIA)—**MISTLETOE FIG**—Code 2
Has smaller leaves. Fruits easily.

F. NERIIFOLIA REGULARIS—**WILLOW LEAVED FICUS**—Code 1
Has narrow 1/4" leaves to 2" long. Will develop aerial roots. Stands reduced light, heat and somewhat dry soil.

F. PUMILA '**MINIMA**'—Code 1
A creeping vine with small leaves. Very slow to develop a trunk. Good for mame bonsai.

F. RETUSA—**BANYAN, INDIAN LAUREL**—Code 1.
Field-grown specimens develop enormous trunks; will develop aerial roots.

F. RUBIGINOSA—**RUSTY-LEAVED FIG**—Code 1
Other *Ficus* with small leaves worth trying include: *F. benjamina* 'Exotica', *F. natalensis, F. philippinensis, F. religiosa, F. celebensis.*

Ficus microcarpa 'Green Island'.

FORTUNELLA HINDSII—HONG KONG KUMQUAT—Code 3

A dwarf tree. Needs cool, dry winters and bright light to produce small oranges (kumquats).

F. MARGARITA—NAGAMI KUMQUAT—Code 3

Culture as above.

FUCHSIA FULGENS—FUCHSIA—Code 3

Needs cool winters. Keep moist when growing. Attractive red flowers hang down. Develops a trunk slowly.

F. MAGELLANICA—HARDY FUCHSIA—Code 3

Culture as above.

GARDENIA JASMINOIDES—GARDENIA—Code 2

Needs acid soil and fertilizer. Provide high humidity and keep evenly moist. Mist frequently. Maintain temperature between 60 degrees F and 70 degrees

F. Has very fragrant white flowers. Cold drafts will cause bud blast or drop.

GREVILLEA ROBUSTA—SILKY OAK—Code 2

Pinnate leaves with orange flowers. Give high humidity and keep moist. Cool winter temperature to 55 degrees F. Bright light. Easy.

G. BANKSII—Code 2

Has red flowers. Culture as above.

GUAIACUM OFFICINALE—LIGNUM VITAE, POCKWOOD TREE—Code 2

Needs very bright light to flower. Has true blue flowers. Keep moist. Needs a slightly acid soil. Very hard, heavy wood. Train by pruning. Branches are hard.

G. SANCTUM—Code 2

Deep green, oblong leaves. Native to Florida. Culture as above.

HEDERA HELIX—**ENGLISH IVY**—Code 3
Very slow trunk thickening. Easily grown in house. Many leaf variations.

HIBISCUS ROSA-SINENSIS—**ROSE-OF-CHINA**—Code 1
Easy to grow. Has large leaves and flowers. Acid fertilizer and bright light induce flowering. Keep moist. Prune after flowering.

H.R–S '**COOPERI**'—**CHINESE HIBISCUS**—Code 1
Has smaller flowers and is variegated. Culture as above.

H. TILIACEUS—**MAHOE**—Code 1
Can take warmer winter temperatures. Culture as above.

ILEX AQUIFOLIUM '**ANGUSTIFOLIA**'—**HOLLY**—Code 3
Narrow leaved English holly. Needs cold winters to 45 degrees F. Difficult.

I. CRENATA '**HELLERI**'—**JAPANESE HOLLY**—Code 3
Conditions and culture as above.

I. CORNUTA—**CHINESE HOLLY**—Code 3
Conditions and culture as above.

I. VOMITORIA '**NANA**'—**DWARF YAUPON HOLLY**— Code 3
Used as an emetic by Indians. Can take somewhat warmer temperatures in winter.

IXORA JAVANICA—**JUNGLE GERANIUM**—Code 1
Has clusters of scarlet flowers. All varieties need acid soil and very bright light to flower. Add iron if leaves become chlorotic. Give high humidity and keep warm.

I. COCCINEA—**FLAME-OF-THE-WOODS**—Code 1
Has red flowers. Culture as above.

JACARANDA ACUTIFOLIA—**FALSE MIMOSA**—Code 1
Has pinnate leaves and showy lavender – blue flowers. Difficult to flower; supply very high light and keep very moist. Flowers are on branch tips, so prune after flowering.

JASMINUM DICHOTOMUM—**PINWHEEL JASMINE**—Code 2
There are many jasmines and most require warm, moist conditions and slightly acid soil. Add chelated iron if they become chlorotic. Most have very fragrant flowers.

JUNIPER CHINENSIS '**NANA**'—Code 3
Needs neutral to slightly acid soil and cool conditions. Very cool rest in winter. Difficult.

J. SQUAMATA '**PROSTRATA**'—Code 3
Conditions and culture as above.

LAGERSTROEMIA INDICA—**CRAPE MYRTLE**—Code 3
Needs high light to flower. Keep moist; treat to prevent fungus. New dwarf varieties are more fungus resistant. Train by pruning. It has tender bark.

LANTANA CAMARA—**LANTANA**—Code 2
Grows outdoors very easily and rapidly in S. Florida. High humidity and bright light induce almost constant flowering. do not allow to go dry. Feed frequently with half-strength fertilizer. Easy.

L. MONTEVIDENSIS—**WEEPING LANTANA**—Code 2
Culture as above.

LAURUS NOBILIS—**LAUREL, BAY**—Code 3
True laurel. Needs cool winter temperatures and bright light.

LEPTOSPERMUM SCOPARIUM—**AUSTRALIAN MYRTLE, NEW ZEALAND TEA TREE**— Code 3
Needlelike leaves; small pink to red flowers like miniature roses. Keep moist, cool and give high light. Prune roots lightly.

LIGUSTRUM JAPONICUM—**JAPANESE PRIVET** —Code 3
Shiny evergreen leaves. Fragrant white flowers. Needs bright light, cool winter temperatures. High summer temperatures require high humidity.

L. J. ROTUNDIFOLIUM—**JAPANESE PRIVET**— Code 3
Culture as above.

L. LUCIDUM—**GLOSSY PRIVET**—Code 3
Has white flowers. Culture as above.

LONICERA NITIDA—**BOX HONEYSUCKLE**— Code 3
Small evergreen leaves; fragrant white flowers. Grows fast. High light and cool winter temperatures needed.

* *MALPIGHIA COCCIGERA*—**MINIATURE HOLLY, SINGAPORE HOLLY**—Code 1
Leaves resemble holly. Light pink flowers and rough, speckled bark. Bright light and warm temperatures help flowering. Keep temperatures above 55° F and do not allow to dry out.

M. GLABRA—**BARBADOS CHERRY**—Code 1
Fruits are edible and very high in vitamin C. It is somewhat weeping in habit. Trunk thickens slowly. Give it slightly acid soil. Recommended.

M. PUNICIFOLIA—Code 1
Similar to *M. glabra* but inedible fruit. Culture as above.

* All malpighias are easily grown and are recommended.

MELALEUCA QUINQUENERVIA—**PAPER BARK TREE**—Code 1
Attractive exfoliating gray–white papery bark. Keep warm, bright light.

MURRAYA PANICULATA—**ORANGE JESSAMINE**—Code 3
Hedge plant in S. Florida. Give high light and cool winter. Flowers are white and fragrant. Water well. Do not allow to dry.

MYRCIARIA CAULIFLORA—**JABOTICABA** — Code 2
Small leaves; brown exfoliating bark. Needs acid soil pH 5.5.-6.0. Fertilize with chelated iron and trace elements. Unusual white flowers and dark purple-blue fruits are produced directly on the trunk and branches. Fruits are sweet and make good jelly. Difficult to flower in a bonsai container. Must be 8-10 years old mimimum. Give very high light.

MYRSINE AFRICANA—**AFRICAN BOXWOOD**—Code 3
Dark green round leaves. Resembles boxwood. Red stems, leaves reduce to one-quarter inch.

MYRTUS COMMUNIS—**MYRTLE**—Code 3
Evergreen leaves with a spicy aromatic scent. Small flowers, blue–black berries. Keep moist. Breaks new shoots from trunk. Cool winter.

M.C. 'MICROPHYLLA'—**DWARF MYRTLE**—Code 3
Small leaved variety. Culture as above.

NOTHOFAGUS CUNNINGHAMII—**TASMANIAN BEECH**—Code 3
Has very small leaves. Keep cool in winter; do not overwater but don't let go dry. Needs good light.

OLEA EUROPAEA—**OLIVE**—Code 3

Old bark is very attractive. Strong bright light, cool winter to 45° F. Let top of soil dry between watering. Takes home temperature to 75° F in summer, but provide high humidity or mist.

OSMANTHUS FRAGRANS—**SWEET OLIVE**—Code 3

Evergreen, fine–toothed glossy leaves. Tiny fragrant cream – colored flowers. Keep cool in winter. Keep moist. Prune to let air flow through.

PINUS ELLIOTTII—**SLASH PINE**—Code 2

Native of S. Florida. Has long needles in bundles of 2 or 3. Needs good drainage. Reduce tap root gradually. New growth breaks on old wood. Needles will reduce with good bonsai culture.

P. HALEPENSIS—**ALEPPO PINE**—Code 3

Native of the Mediterranean basin. It can adapt better than other pines to home heat and dryness. Give bright light, keep slightly moist but do not overwater. Keep cool in winter. Repot every 2 – 3 years. A slow grower.

P. THUNBERGIANA—**JAPANESE BLACK PINE**—Code 3

A two–needle pine. Needs a cold rest period—40°-45° F and cold conditions. Needs bright light and high humidity with good drainage. Many cultivars are available with shorter needles, closer internodes and interesting bark. Very difficult.

PISTACIA TEREBINTHUS—**PISTACHIO, CYPRUS TURPENTINE**—Code 2

Evergreen pinnate leaves. Needs very bright light and cool temperatures. Allow top soil to dry slightly between waterings.

PITHECELLOBIUM FLEXICAULE—**TEXAS EBONY**—Code 2

Light green compound leaves. A slow grower. Can take reduced light and tolerates heat. Keep moist.

P. UNGUIS-CATI—**CAT'S CLAW, BLACK BEAD**—Code 2

Bi–pinnate leaves, small yellow flowers, claw-like seedpods. Prune rather than wire.

PITTOSPORUM TOBIRA—**JAPANESE PITTOSPORUM**—Code 3

Evergreen dense shiny dark green leaves. White fragrant flowers. Needs bright light and cool conditions. Shape by pruning. Keep moist.

PODOCARPUS MACROPHYLLUS MAKI—**BUDDHIST PINE, JAPANESE YEW, SOUTHERN YEW**— Code 3

Needs bright light and good drainage. Leaves like fat needles will reduce, but select for leaf size. Buds break on trunk. Keep evenly moist. High humidity.

P. GRACILIOR— African fern pine—Code 3

Weeping form. Culture as above.

PORTULACARIA AFRA—**ELEPHANT BUSH**—Code 3

Needs good drainage and very high light for compact growth. Pinch regularly. Water moderately. Allow top soil to dry between waterings. Shape by pruning.

PSIDIUM LITTORALE VAR. *LONGIPES*—**STRAWBERRY GUAVA**—Codes 2–3

Interesting mottled exfoliating bark. Evergreen leaves, white flowers and red edible fruit to 1". Water well and then allow surface of soil to dry. Soil slightly acid, pH 5.5. to 6.5.

P. GUAJAVA—**GUAVA**—Code 2-3

Leaves to 4". Culture as above.

PUNICA GRANATUM 'NANA'—DWARF POMEGRANATE—Code 2

Tiny green leaves. Give high light and heat to induce flowering. Blow into flowers to pollinate. Keep moist. Prune after flowering as it blooms on ends of branches. Suckers freely. Easy to grow.

PYRACANTHA ANGUSTIFOLIA— FIRETHORN—Code 3

Attractive white clusters of flowers becoming red-orange berries. They tolerate heat and dryness. Slightly alkaline soil.

P. COCCINEA—SCARLET FIRETHORN

Scarlet fruit. Culture as above.

P. FORTUNEANA

Small leaves and many little red berries. Culture as above.

P. KOIDZUMII

Large berries. Culture as above.

QUERCUS SUBER—CORK OAK—Code 3

From the Mediterranean basin. Needs cool to cold winters. Give high light and humidity. Slow growers. Prune lightly, repot every third year. Difficult.

Q. AGRIFOLIA—CALIFORNIA LIVE OAK

Culture as above.

Q. NIGRA — WATER OAK

Culture as above.

Q. VIRGINIANA—SOUTHERN LIVE OAK

Culture as above.

RAPHIOLEPIS INDICA—INDIAN HAWTHORN —Code 2

Slow growing, has leathery leaves, white to pink flowers and blue–black fruit.

Brittle to wire, needs cool conditions, but stands warmth.

R. UMBELLATA—YEDDO HAWTHORN—Code 2. Culture as above.

RHODODENDRON—AZALEAS—Code 3

Need good light, cool temperature and acid soil. Soil can be from half to all peat moss. All need a cool dormant period in winter. Choose one with small leaf. Bark is thin and easily injured. Use paper-wrapped aluminum wire. Strong bottom growth requires pruning to force top growth.

ROSMARINUS OFFICINALIS—ROSEMARY—Code 3

Evergreen aromatic needlelike leaves, develops a woody trunk slowly. Needs high humidity and cold winters. Keep moist. High light when growing.

R. O. 'PROSTRATUS'—Code 3

A good variety. Culture as above.

SAGERETIA THEA (S. THEEZANS)—POOR MAN'S TEA— Code 2

Mottled bark, small shiny pale green leaves. Grows quickly. Keep evenly moist. Give high light and humidity. Mist frequently in summer.

SEQUOIA SEMPERVIRENS—COAST REDWOOD —Code 3

Feathery fronds, intriguing bark, needs high light, high humidity. Cool to cold winter temperatures. Breaks new buds on older wood. Difficult.

SERISSA FOETIDA—SNOW ROSE, TREE OF A THOUSAND STARS—Code 2

Pinch long, leggy growth constantly. Keep warm, bright light and high humidity. Keep moist; do not let dry. Trunk thickens slowly. Single- and double–flowered forms available. Easy.

SEVERINIA BUXIFOLIA—CHINESE BOX ORANGE—Code 3

Small leaves; needs high light and cool temperatures. Branches are brittle. Wire carefully.

SYZYGIUM PANICULATUM (**EUGENIA PANICULATA**)—**BRUSH CHERRY**—Code 3

Evergreen leaves, creamy colored flowers and rose to purple berries. Needs good light, slightly acid soil. Keep moist. Recommended.

TABERNAEMONTANA DIVARICATA—**CRAPE JASMINE, FLEUR D'AMOUR**—Code 2

Gnarled, gray trunk; glossy green leaves, twisted white flowers. Single and double varieties. keep warm with bright light.

T. DIVARICATA '**CASHMERE**'—**PINWHEEL JASMINE** —Code 2

Pure white pinwheel flowers, can take less light. Culture as above.

TAXODIUM DISTICHUM—**BALD CYPRESS**—Code 3

Give cool dry conditions in winter to rest and induce leaf drop. When growing, water well and give bright light. Difficult.

TRACHELOSPERMUM JASMINOIDES—**STAR JASMINE, CONFEDERATE JASMINE**—Code 1

Has very fragrant white flowers and leathery foliage. Pinch the new growth. Needs cool conditions and high humidity. Allow top soil to dry between waterings.

TRIPHASIA TRIFOLIA—**LIMEBERRY**—Code 1

Keep warm and water well but in winter keep it on the dry side. Bright light. Trunk thickens slowly.

SUGGESTED READING

Compiled by Guest Editor, Sigmund Dreilinger,
with assistance from Brooklyn Botanic Garden librarian,
Deborah Keane

BOOKS

Bonsai: The Complete Guide to Art and Technique. by Paul Lesniewicz. Blandford Press, England. 1984. 192pp. Illust.

Chinese Penjing: Miniature Trees and Landscapes. by Hu Yunhua. Timber Press. Portland, OR. 1987. 171pp. Illust.

The Art of Indoor Bonsai: Cultivating Tropical, Sub-tropical and Tender Bonsai. by John Ainsworth. Trafalgar Square. 989. 128pp. Illust.

The Beginner's Guide to American Bonsai. by Jerald P. Stowell. Kodansha. 1987. 140pp. Illust.

The Creative Art of Bonsai. by Isabelle Samson and Remy Samson. Ward Lock. London. 1988. 168pp. Illust.

Penjing: The Chinese Art of Miniature Gardens. by Hu Yunhua. Timber Press. Portland, OR. 1982. 166pp. Illust.

The Indoor Garden Book. by John Brookes. Crown Pub. NY. 1986. 288 pp. Illust.

Tropical Trees of the Pacific. by Dorothy Hargreaves and Bob Hargreaves. Hargreaves Pub. Hawaii. 1970. 64pp. Illust.

Once Upon a Windowsill: A History of Indoor Plants. by Tovah Martin. Timber Press, Portland, OR. 1989. 300pp. Illust.

The Indoor Light Gardening Book. by George Elbert and Virginia Elbert. Crown Pub. 1973. 250pp. Illust. (Out of print but highly recommended.)

Westcott's Plant Disease Handbook, 4th Ed. by R.K. Horst. Van Nostrand Reinhold. New York. 1979. 803pp.

Diseases and Pests of Ornamental Plants. 5th edition. by Pascal Pirone. Wiley Press. 1978. 566pp.

PERIODICALS

Bonsai: Journal of the American Bonsai Society. Published quarterly by the American Bonsai Society. Keene, NH. Available with Society membership.

Bonsai Australia. Official Journal of the Bonsai Society of Australia. Published quarterly. N.S.W. Australia. (Subscriptions available.)

The Bonsai Bulletin. Published quarterly by the Bonsai Society of Greater New York. Available with a Society membership.

Bonsai Today. Published bi-monthly by Stone Lantern Publishing Company. Sudbury, MA. (Subscriptions available.)

World Tropical Bonsai Forum. Published quarterly by Christine E. Rojas, Inc. Miami, Florida. (Subscriptions available.)

ARTICLES

Donald, Dennis. 1985. *Understanding Soil pH. Bonsai: J. of Am. Bonsai Soc.* 19(4): 14-16.

Stowell, Jerald. 1985. *Tropical bonsai. Bonsai: J. of Am. Bonsai Soc.* 19(4):7-9.

Green, James L. and Bob Rost. 1985. *Many Factors Can Affect Water Quality.* **Amer. Nurseryman.** 161(5): 100-105.

Glass, C. and R. Foster. 1974. *Succulent bonsai.* **Cactus and Succulent Journal.** 46(3): 108-111.

McDonald, Elvin. 1987. *Containers and Potting Mixes.* **BBG Handbook #112: Indoor Gardening.** pp 75-79.

Fitch, Charles Marden. 1987. *Control of Pests and Diseases.* **BBG Handbook #112: Indoor Gardening.** pp 80-81.

PHOTO BY SIGMUND DREILINGER

Buxus microphylla 'Compacta'.